PATHWAYS FOR LOW-CARBON TRANSITION IN BANGLADESH 2025–2050

JANUARY 2025

ASIAN DEVELOPMENT BANK

Contents

Tables and Figures

Tables

Figures

Acknowledgments

The *Pathways for Low-Carbon Transition in Bangladesh 2025–2050* is an in-depth analysis commissioned by the Asian Development Bank (ADB) and developed by the Energy Sector Office, Sectors Group. This report outlines a comprehensive long-term pathway for the country's energy sector and associated greenhouse gas emissions, focusing on low-carbon energy technologies and efficiency measures. It employs a rigorous modeling approach utilizing the MESSAGE-ix energy model developed by the International Institute for Applied Systems Analysis.

The study was carried out under the leadership of Kee-Yung Nam, principal energy economist, with supervision from Priyantha Wijayatunga, senior director of the Energy Sector Office, Sectors Group. The report was prepared and the modeling was conducted by the management and staff of PricewaterhouseCoopers International Limited. David Morgado, senior energy specialist, coordinated and reviewed the content. John Weiss, serving as the technical editor, synthesized the three documents into an integrated report.

A team of both international and national experts offered significant insights as peer reviewers. Maria Dona Aliboso, associate project officer, and Maria Carmela Abadeza, associate project analyst, managed the production process and provided analytical and administrative support. Ma. Theresa Mercado copyedited the report, while Lumina Datamatics did the layout. Jess Macasaet, proofreader and page proof checker also contributed to the production of this publication. The study also gained valuable insights and feedback from colleagues at the ADB Energy Sector Office, Sectors Group.

This study is supported by funding from the Clean Energy Fund under the Clean Energy Financing Partnership Facility, the People's Republic of China Poverty Reduction and Regional Cooperation Fund, the Urban Climate Change Resilience Fund, and the E-Asia Knowledge Partnership Fund.

Abbreviations

ADB	Asian Development Bank
BAU	business-as-usual
BCSIR	Bangladesh Council of Scientific and Industrial Research
BEEER	Building Energy Efficiency & Environment Rating
BNBC	Bangladesh National Building Code
BSEC	Bangladesh Securities and Exchange Commission
CAGR	compound annual growth rate
CAP	Country Action Plan
CC	Combined Cycle
CCS	carbon capture and storage
CCU	carbon capture and utilization
CCUS	carbon capture, utilization, and storage
CDM	Clean Development Mechanism
CNG	compressed natural gas
CO_2	carbon dioxide
CSR	corporate social responsibility
DC	direct current
EECMP	Energy Efficiency and Conservation Master Plan
EE&C	energy efficiency and conservation
ESCOs	energy-savings companies
FDI	foreign direct investment
FY	fiscal year
GDP	gross domestic product
GHG	greenhouse gas
GIZ	German Development Agency
GTF	Green Transformation Fund
H_2	hydrogen
HFC	hydrogen fuel cell
HVAC	heating, ventilation, and air-conditioning
IDCOL	Infrastructure Development Company Limited
IEA	International Energy Agency

IFAD	International Fund for Agricultural Development
IFC	International Finance Corporation
IGCC	integrated gasification combined cycle
IIASA	International Institute for Applied Systems Analysis
INDC	Intended Nationally Determined Contribution
IPP	independent power producer
JICA	Japan International Cooperation Agency
LED	light-emitting diode
LEED	Leadership in Energy and Environment Design
LNG	liquefied natural gas
LPG	liquefied petroleum gas
MAC	marginal abatement cost
MDB	multilateral development bank
MEPS	minimum energy performance standards
MoEFCC	Ministry of Environment, Forests and Climate Change
MRT	mass rapid transit
NDC	nationally determined contribution
P2P	peer to peer
PHEV	plug-in hybrid electric vehicle
PHS	pumped hydro storage
PPI	private participation in infrastructure
PPP	public–private partnership
PV	photovoltaic
RAJUK	Radjhani Unnayan Kartripakkha
RESA	Renewable Energy for Smallholder Agriculture
SDG	Sustainable Development Goal
SFD	Sustainable Finance Department
SMR	steam methane reformation
SREDA	Sustainable and Renewable Energy Development Authority
SSP	shared socioeconomic pathway
STP	strategic transport plan
SURE	Scaling Up Renewable Energy
TFED	total final energy demand
TPES	total primary energy supply
TRL	technology readiness level
UNFCCC	United Nations Framework Convention on Climate Change
US	United States
USAID	United States Agency for International Development
USC	Ultra-supercritical

Weights and Measures

°C	degree Celsius
GW	gigawatt
GWh	gigawatt-hours
kg	kilogram
kWh	kilowatt-hour
MMSCFD	million standard cubic feet per day
Mtoe	million tons of oil equivalent
$MtCO_2$	million tons of carbon dioxide
$MtCO_2e$	million tons of carbon dioxide equivalent
MWe	megawatt electric
tCO_2	tons of carbon dioxide
TWh	terawatt-hour
TR	tons of refrigeration

Executive Summary

Bangladesh has developed rapidly over the past decades. However, the development of the renewable energy supply (mainly hydro and biofuels) has not kept pace with the needs of its economy, which has grown to rely on higher carbon energy (mainly coal and oil). In 1990, over 62% of the primary energy was supplied from clean energy sources. By 2019, this situation had changed drastically with more than 55% of the primary energy being supplied from fossil-based energy sources.

Emissions from energy-related sources have more than quadrupled since 2000 from 20.9 million tons of carbon dioxide ($MtCO_2$) to 90.7 $MtCO_2$ in 2021 (World Bank 2021). This can be attributed to the rising energy demand due to population and industrial growth coupled with progress on the Sustainable Development Goal (SDG) agenda to provide adequate access to affordable, reliable, sustainable, and modern energy to all (SDG 7). Energy consumption is expected to rise considerably in the future with the rise in population and economic development. The economy is anticipated to more than double between 2020 and 2050. Also, the country's population is expected to increase by 29 million people between 2020 and 2050. Further economic growth and increasing urbanization will lead to changes in consumer preferences, behavior, and lifestyle changes resulting in higher energy demand.

The Paris Agreement, adopted at the 21st meetings of the Conference of Parties to the United Nations Framework Convention on Climate Change (UNFCCC) toward combating climate change, calls for global efforts in form of greenhouse gas (GHG) emission reduction commitments and climate change adaptation from all signatory countries. It further requires nations to strengthen their mitigation-related commitments over time.

The Government of Bangladesh submitted its Intended Nationally Determined Contribution (INDC) to UNFCCC on 25 September 2015, for three sectors: power, industry, and transport. In the INDC, the government proposed a 5% unconditional reduction in GHG emissions relative to the base year of 2011 by 2030 and a further 10% conditional reduction conditional on support from the international community. In August 2021, the INDC was updated to include additional sectors and enhanced its ambition for mitigation.

This study commissioned by the Asian Development Bank (ADB) explores future pathways for the energy sector and national GHG emissions up to 2050. The widely used MESSAGEix energy model has been applied with a scenario-based approach contrasting two alternative scenarios. A business-as-usual (BAU) scenario assumes continuation of current and historic policy and technology trends and a low-carbon scenario incorporates policy shifts and technical change in support of the Paris Agreement goals. Both scenarios use the same projections for population and gross domestic product (GDP) per capita (derived from Shared Socioeconomic Pathways [SSP2]). The model shows that a reduction in energy use and associated emissions is achievable with a departure from current trends and policies; however, while both scenarios have a high investment cost, the low-carbon pathway is more expensive and requires significant behavioral change.

Energy Efficiency

The study identifies improvements in energy efficiency, electrification, and fuel-switching to low-carbon fuels as the three main pillars of a decarbonization strategy for Bangladesh. With the increasing adoption of energy-efficient technologies, processes, and appliances, built into the low-carbon scenario, projected final energy requirements fall over time resulting in energy and emissions savings. Additionally, from the supply side, a switch to more efficient power production processes also contribute to the reduction in energy demand. Total final energy demand (TFED) is estimated to grow from 42.1 million tons of oil equivalent (Mtoe) in 2020 to approximately 92.2 Mtoe by 2050 under the BAU scenario, and to 87.8 Mtoe by 2050 under the low-carbon scenario, an overall reduction of approximately 5%. Of the total energy demand reduction of 4.4 Mtoe in 2050, the majority (almost 87%) is from the residential (cooking) sector, followed by 17% from industry and 4% from agriculture. Critical here is the successful adoption of energy-efficient cooking processes and a shift away from traditional biomass-based cooking. Demand increases in transport due to higher traffic flows, with a share in reduction relative to the BAU of −8%.

Electrification

Electrification is a potent energy transition strategy for the transport and residential cooking sectors. The low-carbon scenario assumes a rapid spread of electric vehicles (and the associated infrastructure required for their use) and electricity-based cooking appliances. This increases overall national electricity demand. To meet this demand by 2050, electricity generation under the low-carbon scenario reaches 516 terawatt-hours (TWh), a 32-TWh increase over generation of 484 TWh in the BAU scenario. However, for electrification to be a successful strategy for decarbonization, it is essential to move away from fossil-based power generation to that based on renewables. Hence the low-carbon scenario requires a phased transition from coal to renewables, like solar and wind. In 2050 under the low-carbon scenario, solar photovoltaic (PV) contributes 15% of generation and onshore wind 2%. Gas remains the major fuel for power generation under both scenarios, 49% in the low-carbon scenario and 64% in the BAU, since with significant additions of renewable-based power capacity, gas is essential to cover variations in renewable supply and to meet peak needs. Given that proven reserves have only a limited life and are low relative to current annual consumption, discovery and verification of new gas fields is a priority. The low-carbon scenario also involves more generation supplied by low-carbon electricity imports (13%) and nuclear-based power (8%) than in the BAU case, where the shares are 2% and 4%, respectively. On the other hand, a critical difference is that coal-based electricity generation is virtually phased out in the low-carbon scenario (at 1% of generation), whereas it still takes 22% of generation under the BAU. Even though all new coal plants are taken as using the more efficient supercritical technology, this still creates a significant difference in emissions between the two scenarios.

Total primary energy supply (TPES) is projected to grow from 52.4 Mtoe in 2020 to 141.2 Mtoe in 2050 under the BAU scenario and to 127.8 Mtoe in 2050 under the low-carbon scenario. Natural gas takes around 60% of TPES under both scenarios. Apart from the lower requirement for energy, the low-carbon scenario has a significant shift in fuels toward cleaner alternatives, like renewables and natural gas. Under the low-carbon scenario, around 8% of the TPES in 2050 comes from solar and wind power, as compared with virtually zero under the BAU scenario. Under the BAU coal (both domestic and imports) still takes around 19% of TPES, whereas domestic coal does not figure in the low-carbon scenario. The efficiency gains achieved through process efficiency improvements, electrification, and fuel-switching, as envisaged under the low-carbon scenario, cause total energy imports in 2050 to be 48.1 Mtoe as compared with 64.2 Mtoe in the BAU scenario. Natural gas becomes the major fuel imported in the low-carbon scenario. This is primarily because the energy transition in a range of sectors from coal, biomass, and oil to natural gas is predicted to lead to a growth in natural gas imports. Once this occurs on a large scale, the government will need plan carefully to minimize the geopolitical risks to the country's energy security.

Dhaka City

Since Dhaka is an important part of the clean energy transition a similar analysis to that conducted nationally is applied for the city area. Under the BAU scenario, TFED for Dhaka is estimated to grow from 2.53 Mtoe in 2020 to approximately 9.62 Mtoe by 2050. With measures for efficient utilization of energy, in the low-carbon scenario in 2050 demand is predicted to be 8% lower at 8.81 Mtoe. Out of the total energy demand reduction relative to the BAU scenario in 2050 of 0.81 Mtoe, 0.97 Mtoe would be from industry, followed by 0.55 Mtoe from residential cooking. However, an increase of 0.49 Mtoe relative to the BAU is projected in the commercial sector and of 0.23 Mtoe in the transport (passenger) sector due to higher urbanization and the increase in vehicles in the city. The reduction in final energy demand will also influence the primary energy requirement of the city. The TPES is estimated to reach 10.76 Mtoe in 2050 under the BAU, and to be 6% lower in the low-carbon scenario. While the overall primary energy supply remains relatively similar, there is a significant change in the energy supply mix in the low-carbon scenario with a rise in the share of solar PV and a larger share of cleaner-carbon electricity imports from the national grid, and a fall in the share of natural gas. The combined share of solar PV and solar thermal increases from 3% in 2020 to 12% in 2050, predominantly solar PV off-grid applications across the residential, commercial, and industry sectors.

Emissions

The application of best-practice low-carbon technology for new investment is built into both scenarios, although the speed and degree of their uptake differs. This involves the use of combined cycle gas boilers, electric vehicles, electric cooking appliances, new building standards, electric irrigation pumps and so forth, and the infrastructure necessary to support them. Only technologies currently available commercially are considered, and green hydrogen-based technologies, for example, with potential for use in transport and industry, are not incorporated in the model. With rising economic activity, both scenarios see rising emissions from 2020, but in the low-carbon scenario, these are 16% lower in 2030 and 38% lower in 2050 relative to the BAU scenario. This allows the short-term nationally determined contribution (NDC) target for emissions to be met. In the longer term the difference between the two scenarios in 2050 is 134.9 $MtCO_2e$ of emissions. This difference is due primarily to emissions reduction on the energy supply side (83%), followed by 8% due to efficiency improvements in industry and 7% due to improvements in residential cooking, because of the shift away from cooking with traditional biomass to that based on electricity and gas.

Estimates of the cost-effectiveness of different forms of emissions mitigation give a wide range of values. However, some forms of mitigation are less cost-effective than others. Onshore wind power is considerably cheaper than offshore wind, grid-connected solar is cheaper than off-grid solar, electric pumps are cheaper than off-grid solar pumps, and supercritical coal technology is a considerably more expensive form of mitigation compared with natural gas-based technology. Overall, on a weighted basis, allowing for differences within the categories, natural gas is the most cost-effective fuel and solar power is cheaper than wind power.

The picture is similar for Dhaka. At present, the city depends almost entirely on imported power from power stations that are part of the national grid but located outside the city. Only a very small amount of power consumed in the city is produced there from solar off-grid applications. Under the low-carbon scenario, grid-connected solar PV is expected to remain a small share of the total electricity mix due to land constraints in the city hindering the development of solar parks; however, solar off-grid installed capacity is expected to rise. Comparing the two scenarios there is a total emissions reduction of 40% in the low-carbon scenario as compared with the BAU scenario. The reduction in emissions relative to BAU is primarily due to energy efficiency interventions in the demand sectors. Just over 50% of the reduction is due to improvements in transport with the assumed shift to electric vehicles, 30% to improvements in residential cooking with the move away for traditional methods, and 16% to improvements in industry and construction. The model suggests that under the low-carbon pathway by 2050, the city will take roughly 10% of national total final demand (8.81 Mtoe out of 115.8 Mtoe), and 10% of total national emissions (22.8 $MtCO_2e$ out of 222 $MtCO_2e$). However, its share in demand-side emissions is much higher at around 20% (22.6 $MtCO_2e$ out of 115.8 $MtCO_2e$). This high proportion is likely to be due to the concentration of residential buildings and related activity and the extensive transport links in the city.

Investment Requirements

Funding of the low-carbon transition is expensive. Investment under the BAU scenario is estimated for three periods: 2020–2030, 2030–2040, and 2040–2050. The costs are $60 billion (2020–2030), $57.3 billion (2030–2040), and $41.8 billion (2040–2050) in constant 2010 US dollars. Hence funding in current prices will be much higher. Under the BAU scenario, major investments in power would be required for coal supercritical, gas combined cycle, and nuclear power generation. Simultaneously, a significant investment is also required for upgrading and expanding the electricity grid. Within demand sectors, major investment needs also arise from the residential and transport sectors aimed at their electrification. By 2050, some investment for the development of solar PV grid and off-grid technologies are also included in the BAU scenario.

Total investment needs under the low-carbon scenario are estimated to be $61.3 billion (2020–2030), $64.9 billion (2030–2040), and $105.7 billion (2040–2050), again in constant 2010 prices, which is an increase of 45% over the BAU scenario. Major investments are expected in solar, wind, and biomass technologies for decarbonization of the energy supply. Another major difference is the much larger investment required in the domestic gas and off-grid solar systems used across the agriculture, residential, and industry sectors. As in the BAU scenario, the majority of the demand-side investment is expected in the transport and residential sectors, particularly targeting electrification measures. Overall, in 2020–2050, $73 billion more investment above that in the BAU scenario is needed. This is in 2010 constant prices, so in current prices, requirements are much higher. Estimated investment requirements for Dhaka and the national total are not comparable, as they are in different sets of prices.

Policy Implications

Finance for the required investment will need to draw heavily on foreign sources—both concessional climate finance and foreign direct investment (FDI) involving public–private partnerships (PPP). In addition, there is a need to involve the domestic private sector. Public funding can help to a limited degree to stimulate this process, but realistically the scope for this will be limited. While the pathway set out in the low-carbon scenario shows what can be achieved and its investment cost, delivery of this pathway will be dependent on the success of policy interventions and access to finance. The discovery of new gas fields and the expansion of existing ones will also be essential given gas' continued importance as a fuel in the low-carbon scenario.

Best available technologies must also be deployed across different sectors. However, adapting and modifying these to local conditions will be important. Research and development support is crucial for fostering technological innovation, which when complemented with standards, certifications, and codes offers important tools for a decarbonizing strategy. Economic incentives and subsidies can be utilized to nudge consumer and producer behavior toward cleaner alternatives and the careful use of resources. For this to happen prices in the market must incorporate environmental costs, so that producers and consumers can adequately take account of environmental impact in their decisions on what to produce and consume. As the role of financing is critical, the involvement of the private sector requires that the investment climate is supportive and that the incentive package offer is competitive.

A strong legal and regulatory framework to support legal and regulatory interventions, such as the mandatory enforcement of minimum energy performance standards (MEPS) for various appliances in residential and commercial buildings, accompanied by stringent monitoring and evaluation of progress on achieving a low-carbon targets is central. High standard and coherent institutions are needed. Institutional coordination is important so that, for example, energy and climate ministries can identify and resolve differences between sectoral priorities and policies, and jointly support environmental initiatives. Similarly, national and city-level institutions should coordinate on implementing the necessary policy interventions; for example, the monitoring, inspection and enforcement of the Bangladesh National Building Codes (BNBC) could be embedded within the mandates of city corporations.

Finally, public support and engagement will be needed. It is necessary to increase public understanding and generate awareness about the available low-carbon options and the policies and measures aimed at their adoption. Behavioral changes, like a preference for nonmotorized transport, a willingness to conserve energy, and a modal shift toward public transport, can lead to a reduction in energy demand. However, consumers must be convinced of the merits of such changes and be made aware of their consequences in terms of environmental impact.

By identifying the technological measures required on the supply and demand sides to accelerate decarbonization and their cost, this road map is intended to inform long-term energy planning and to show governments, investors, households, and other stakeholders what meeting the national climate commitments means in terms of production, technology, and investment choices. If the measures identified in the low-carbon scenario are implemented, then the model suggests long-term national climate targets can be achieved, at the same time as economic growth rates are maintained.

1. Model and Its Technologies

Introduction

Bangladesh is currently one of the fastest-growing economies in South Asia, but with limited fossil fuel reserves. Over 90% of the country's power-producing thermal plants are gas-based, and gas is also needed for the industrial sector (Pranti et al. 2013). The government has ambitions to raise GDP per capita to middle-income status and energy will play a vital role in future growth. The country will need an effective strategy to ensure a low-carbon development, while also addressing the barriers to higher growth posed by low access to reliable and affordable power, limited availability of usable land, rapid urbanization, and vulnerability to climate change and disasters triggered by natural hazards (PricewaterhouseCoopers 2018).

In relation to the Paris Agreement climate targets in the first round of NDCs, the government pledged an unconditional target of 5% emissions reduction below BAU GHG emissions (2011 base year) by 2030 and a further conditional target of 10% reduction (Ministry of Environment, Forests and Climate Change [MoEFCC] 2015). The updated NDC (2021) proposed an unconditional 6.7% reduction below BAU GHG emissions by 2030 and an additional 15.1% reduction in the conditional scenario (MoEFCC 2021). Additionally, the government also aims to reduce energy intensity (national primary energy consumption per GDP) in 2030 by 20% compared to the 2013 level (Energy Efficiency and Conservation Master Plan (EECMP) 2016).

Methodology and Assumptions

The model used for this study is the MESSAGEix modeling framework developed by the International Institute for Applied Systems Analysis (IIASA). MESSAGEix is a versatile, open-source, dynamic systems-optimization model providing a framework for analyzing an energy system with its interdependencies, allowing for resource endowments and resource potential, extraction rates, imports, exports, the generation of electricity, and the conversion of fuels to energy for end-use. The optimization model gives the least-cost solution subject to constraints over predefined time periods.

The MESSAGEix model is used in a range of countries for the assessment of national and subnational strategies for energy transition. It has the flexibility to evaluate and assess the mitigation potential of sectors under various low-carbon development scenarios and allows the assessment of the impact of different policy and technology choices both nationally and across sectors. The five sectors considered for this study are agriculture, commercial, industry, residential, and transport. To establish the pathways for energy transition, two scenarios are used:

- *Business-as-usual scenario*—assumes continuation of past and current trends into the future. This scenario derives future patterns in technological penetration and resource consumption based on historical trends.
- *Low-carbon scenario*—assesses options for achieving energy and emissions reduction trajectories based on the assumption of enhanced ambition for low-carbon development in support of goals of the Paris Agreement.

These scenarios are not purely cost-optimization scenarios as a level of technological guidance has been provided to maintain consistency with current and historic trends in the case of the BAU scenario and for decarbonization policy initiatives in the case of the low-carbon scenario. The general framework in each of the scenarios is same. However, parameters can be defined exogenously in accordance with the policy framework underlying each scenario, such as the assumptions regarding degree of policy and behavioral change, technological cost curves, technological efficiency improvements, or resource availability. The modeling time frame considered spans 2020 to 2050 with data given at five-year intervals. The model is applied nationally and for the capital city of Dhaka (see Section 5: Low-Carbon Pathway for Dhaka City).

Business-as-Usual Scenario

The BAU scenario projects future energy use and emissions based on cost minimization, while allowing for historic trends in installed capacities and utilization, resource potential, and growth in end-use energy demand from key economic sectors, covering agriculture, commercial, residential, industry, and transport. Projections for use of energy commodities such as coal, oil, gas, biomass, and hydro are based primarily on current and historic utilization patterns. The uptake of emerging renewables and cleaner technologies such as solar, wind, and nuclear is guided by cost without any forced constraint imposed by environmental policy. A limitation of the analysis of both scenarios is that for transport the study only considers registered vehicles as reported in the relevant national statistical publications of Bangladesh. Accounting for unregistered vehicles, especially motorized ones, would provide a more reliable estimate of energy use in the transport sector.

Low-Carbon Scenario

The low-carbon scenario takes into consideration a set of policy and technology measures on both the supply and demand sides to steer the economy toward a low-carbon pathway. These measures include demand-side electrification, supply and demand-side efficiency improvement, and fuel-switching across supply and demand sectors toward carbon-neutral or low-carbon fuels, such as natural gas, biofuels, and off-grid renewable energy generation. The INDC targets are not explicitly provided as an input, rather the scenario outcomes are evaluated to investigate the potential impact of policy interventions in progressing toward the INDC targets.

Some demand-side management and behavioral interventions assumed in the low-carbon scenario are reflected at the energy or end-use level. The final to end-use energy conversion efficiencies used reflect the process and technological efficiency judged to be achievable over the time periods involved. Key targets applied in the low-carbon scenario with a view to meeting the country's international commitment to emissions reduction to address climate change are in Table 1. The modeling time frame considered is from 2020 to 2050, with data given at five-year intervals. These scenarios are not purely cost-optimization scenarios, as assumptions about policies and technologies in use are imposed as constraints. In the BAU scenario, this is to maintain consistency with current and historic trends and in the low-carbon scenario to maintain consistency with the low-carbon policy initiatives and associated targets. This implies that in the low-carbon scenario costs are minimized, subject to imposed constraints in the form of targets as specified in Table 1. GDP per capita and population growth assumptions used in the model are the same in both scenarios. These are taken from the Shared Socioeconomic Pathways (SSPs), which are the most recent version of consistent global population, urbanization, and GDP projections used to develop global emissions projections (IIASA 2018). The model uses the mid-range SSP2 case as the basis for the scenarios. The assumptions on improvements in technological energy efficiencies, aligned with current policies and trends, are common across BAU and low-carbon scenarios. Additional energy efficiency improvements under the low-carbon scenario result from strategies like electrification and fuel-switching.

Table 2 and Table 3 provide an overview of socioeconomic parameters from the SSP2 scenario for Bangladesh (IIASA 2018).

Table 1	Summary of Low-Carbon Targets Adopted in the Low-Carbon Scenario
Scenario	Target description
INDC	Reduce 6.73% (unconditional) and further 15.12% (conditional) emissions compared to baseline by 2030 as per updated INDCs
Fuel-shift	40% share of liquefied petroleum gas (LPG)-based residential cooking at final energy level by 2030
	More than 15% share of electric vehicles in roadways and railways at final energy level by 2035
	Achievement of more than 25% use of solar-based pumps at final energy level by 2030
Renewable Energy	4 gigawatts (GW) of wind power capacity by 2030
	25 GW of total solar PV capacity by 2030
Energy Efficiency	20% reduction in energy intensity of GDP compared to 2013 levels by 2030
	50% light-emitting diode (LED) light penetration in the residential and commercial sectors by 2030 and 90% by 2050

Source: Compiled by authors.

Table 2	Population Projections (millions)						
2020	2025	2030	2035	2040	2045	2050
166.2	174.3	181	186.7	191	194.1	195.7

PPP = purchasing power parity.
Source: Riahi et al. 2017.

Table 3	Gross Domestic Product (PPP) Projections ($ billion 2010/yr)						
2020	2025	2030	2035	2040	2045	2050
719	835	961	1,106	1,282	1,485	1,717

PPP = purchasing power parity.
Source: Riahi et al. 2017.

Technologies Incorporated in the Model

The model developed for the analysis considered various technologies with significant potential for decarbonizing the economy. Based on a literature review and stakeholder consultation, a list of key technologies was developed (Table 4). The appendix discusses the future potential of hydrogen-based applications for low-carbon growth but does not include them in the low-carbon scenario as the commercial readiness of green hydrogen (H_2) is not yet adequate and there is no clarity on policy or plans for the development of green H_2 in the country. Similarly, carbon capture utilization and storage (CCUS) is also discussed but given its still relatively low technology readiness level (TRL), this technology is also omitted from the model.

Table 4	Key Technologies for Low-Carbon Road Map		
Technology	Description	Country context	Policy link
Coal Ultra-supercritical (USC)	Clean coal technology requires less coal per megawatt-hour (MWh) and operates at an efficiency of higher than 45% thereby additionally reducing operating costs.	Bangladesh is implementing USC technology in Matarbari 1,200 megawatts (MW), Kohelia 700 MW, Sumitomo 1,200 MW, and coal-fired power plant (CPGCBL 2019).	
Coal Supercritical Power Plant	Clean coal technology requires less coal per MWh and operates at an efficiency of higher than 42% compared to standard coal-fired plants that operate at around 33%.		INDC (2021) implementation road map—100% of new coal-based power plants to use supercritical technology.
Coal Subcritical Power Plant	A conventional process where coal is processed under relatively lower pressure and temperature conditions. Costs are low; however, emissions are higher, and efficiency is significantly lower at around 33%.		
Gas Combined Cycle (CC) Power Plant	The combined cycle process involves extracting heat from both the input and exhaust stream. This helps reduce energy wastage, reduce emissions, and increase efficiency to over 60% compared to single-cycle efficiency of 34%.	Gas CC power plant of capacity 450 MW in Meghnaghat. The ADB has provided for the financing a of 718 MW plant to be set up by Reliance Power in collaboration with JERA and General Electric.	
Energy Recovery from Liquefied Natural Gas (LNG) Regasification	The process through which losses during the LNG regasification process can be recovered and used as a useful cooling source. The recovered energy can be used as electricity, cooling (for cold chain and/or refrigerated trucks), and air-conditioning.	Bangladesh has two LNG terminals commissioned. The current send-out capacity of the country is at 3,700 million standard cubic feet per day MMSCFD of natural gas.	
Coal Integrated Gasification Combined Cycle (IGCC)	Next-generation power generation system with enhanced efficiency and environmental performance due to the combination of coal gasification and gas turbine combined cycle.	CPGCBL and Mitsui & Co Limited, Japan seeks to develop a jointly imported LNG-based gas-fired combined cycle power plant with a capacity of 500–600 MW capacity (Mitsubishi Hitachi Power Systems, n.d.).[a] Another combined cycle power plant is Haripur (412 MW). Upcoming projects from EGCB are Siddhirganj (335 MW), Munshiganj Combined Cycle Power Plant project (Phase I, 660 MW and Phase II, 660 MW), Feni CCPP (Phase I, 660 MW; Phase II, 660 MW; Phase III, 660 MW; Phase IV, 660 MW).	

continued on next page

Table 4	*continued*		
Technology	**Description**	**Country context**	**Policy link**
Diesel Generator	Combination of diesel engine and electric generator to produce electricity. Typically, only used as a backup in places without grid connectivity as fuel efficiency is low.		
Pumped Hydro Storage (PHS)	PHS stores electrical energy as the potential energy of water. Generally, this involves pumping water into a large reservoir at a high elevation—usually located on the top of a mountain or hill.	Bangladesh has a low dependency on hydroelectric power projects, with the only station being Kaptai at 230 MW. However, the governments of Bangladesh and Nepal have agreed to build two pumped-storage hydropower plants with a capacity of more than 1,600 MW in Nepal. Electricity produced will be exported to Bangladesh via India.	
Nuclear Power Plant	A facility that converts atomic energy into usable power. The heat produced by nuclear fission in a reactor is used to drive a turbine to power electric generators.	Nuclear power has recently gained prominence in Bangladesh with two reactors currently under construction in Rooppur, expected to start commercial operations by 2023–2024 with a capacity of 1,200 megawatt-electrical (MWe) each.	Nuclear Power Plant Act (2015)
Wind Power Plant	The wind is used to drive mechanical turbines to generate power. Offshore set up typically has greater power output and lesser ecological impacts; however, it has higher investment and maintenance costs compared to onshore plants.	Wind energy potential in Bangladesh is estimated at over 34 GW. Most of the country has low wind speeds with wind energy sites concentrated in few pockets. However, with newer technology, it is now possible to tap into regions of low wind speed as well. Recently government approved a proposal to construct the first wind power plant at Mogla having a capacity of 55 MW.	There is no dedicated policy on offshore wind power development in the country currently.
Biomass Power Plant	Wood waste or other waste is burned to produce steam that runs a turbine to make electricity. With pollution controls and combustion engineering net emissions from biomass are less than conventional fossil fuels.	Infrastructure Development Company Limited (IDCOL) produces 300 kilowatts (kW) power from biomass and the government power plant produces 1 MW power from the biomass-based system. However, biomass is primarily used in rural areas for heating and cooking purposes. The government is aiming for the establishment of waste to energy power plants, with two planned plants at landfills near Dhaka expected to generate 35 MW each.	Renewable Energy Policy (2015)

continued on next page

Table 4	*continued*		
Technology	**Description**	**Country context**	**Policy link**
Solar Photovoltaic (PV) Grid	PV panels or arrays are connected to the utility grid through a power inverter unit allowing them to operate in parallel with the electric utility grid. Extra power generated can be stored in batteries or feedback into the grid.	It is estimated that Bangladesh receives around 1,900 kilowatt-hours (kWh)/square meter (m²) per year. The current installed solar capacity is 416 MW. A 73-MW solar park in Mymensingh started commercial operations in 2021. Plans to set up a solar plant of 100-MW capacity in Jamalpur have been finalized. The government is aiming for 40 GW of installed solar capacity by 2041. The National Solar Energy Roadmap under high deployment scenario anticipates future solar capacity of 30 GW of which 40% would be covered by large-scale solar PV and around 40% by rooftop solar PV systems.	Draft National Solar Energy Roadmap (2020); Renewable Energy Policy (2015); Draft Renewable Energy Policy (2022)
Solar-powered Irrigation Systems (Solar Irrigation Pumps)	Agricultural pumps use solar power instead of electricity or diesel.	The World Bank has signed a $55-million financing agreement with the government to support the installation of 1,000 solar irrigation pumps, 30 solar mini grids, and about 4 million improved cookstoves in rural areas (PVTECH, 2018).	Target to scale up the potential of solar irrigation pumps—Renewable Energy Development Target National Renewable Energy Policy 2008 Power System Master Plan 2016
Floating Solar	Floating solar PV is a solar PV application in which PV panels are designed and installed to float on water bodies such as reservoirs, hydroelectric dams, industrial ponds, water treatment ponds, mining ponds, lakes, and lagoons.	The government has set a target to generate 200 MW from floating solar in 5 years starting in 2019. The rationale being the scarcity of land for ground-mounted solar projects is forcing consideration of this as an alternative option. The first project funded by ADB has commenced at Mongla with an installed capacity of 10 kW. The project is set to be scaled up to 15 MW based on feasibility.	Draft National Solar Energy Roadmap (2020); Renewable Energy Policy (2015)
Decentralized Generation and Mini Grids	A mini grid is defined as a system having a renewable energy-based electricity generator (with a capacity of 10 KW and above), and supplying electricity to a target set of consumers (residents for household usage, commercial, productive, industrial, institutional setups, etc.) through a public distribution network.	World Bank has signed a $55-million financing agreement with the government to support installation of 1,000 solar irrigation pumps, 30 solar mini grids, and about 4 million improved cookstoves in rural areas.[b] The Second Rural Electrification and Renewable Energy Development Project has 10 solar mini-grids in remote areas.	Solar Homes Program

continued on next page

Table 4	*continued*		
Technology	**Description**	**Country context**	**Policy link**
Battery Energy Storage Systems	Energy storage could apply to different technologies ranging from PHS, flywheels, supercapacitors, compressed air, thermal energy storage, and batteries. Advanced energy storage technologies are capable of dispatching electricity within seconds and can provide power backup ranging from minutes to many hours.	A 36-kW-peak solar PV-battery storage power plant pilot project has been developed at Siddhirganj through ADB financing.[c]	
Electric Vehicles	Electricity-based technology that would become an alternative to internal combustion engine vehicles	The energy demand from the transport sector is set to increase to 12.2 million tons of oil equivalent (Mtoe) in 2040, out of which more than 60% is from road transport, which is set to use oil as its primary fuel. To reduce the dependence on oil and reduce associated costs, electric vehicles could be considered as an option. This shift is slower in Bangladesh compared to India and the People's Republic of China. Presently, the government is registering 100,000 rural electric vehicles to initiate the penetration of electric vehicles in the transport mix.	PERSPECTIVE PLAN OF BANGLADESH 2041 formulated in 2018, 8th Five-Year Plan (2020–2025)
High-Efficiency Residential and Commercial Heating, Ventilation, and Air-Conditioning (HVAC)	Use of inverter technology and fixed speed air-conditioners for better performance of the air-conditioners along with common metric and standards and labeling program to increase savings and quality.	As the cooling demand in the country is increasing, the market for energy-efficient air-conditioners will increase to curb the rising energy use for cooling.	Energy efficiency labeling program, BNBC, Leadership in Energy and Environment Design (LEED) Certification, The Building Energy Efficiency & Environment Rating (BEEER)
District Heating and Cooling Systems	A system that can distribute heating and cooling generated in a centralized location through a network of pipes to communities and other establishments.	As the cooling demand for the country is increasing manifold district systems coupled with renewable energy sources for powering these systems can be looked upon as a viable solution.	BNBC, LEED Certification, The BEEER.

[a] Daily-sun, 2017. CPGCBL, Mitsui sign deal for 600MW LNG power plant, published on 8 November 2017, accessed online on 23 November 2024 at https://www.daily-sun.com/post/267171.

[b] S. Prateek. 2018. World Bank Approves $55 Million for Renewable Energy in Bangladesh. 16 April. Mercom.

[c] Improving Lives of Rural Communities Through Developing Small Hybrid Renewable Energy Systems, 2017.

Source: Compiled by authors.

2. National Results of the Model

To estimate the energy demand for each sector, the following approach was adopted (Figure 1).

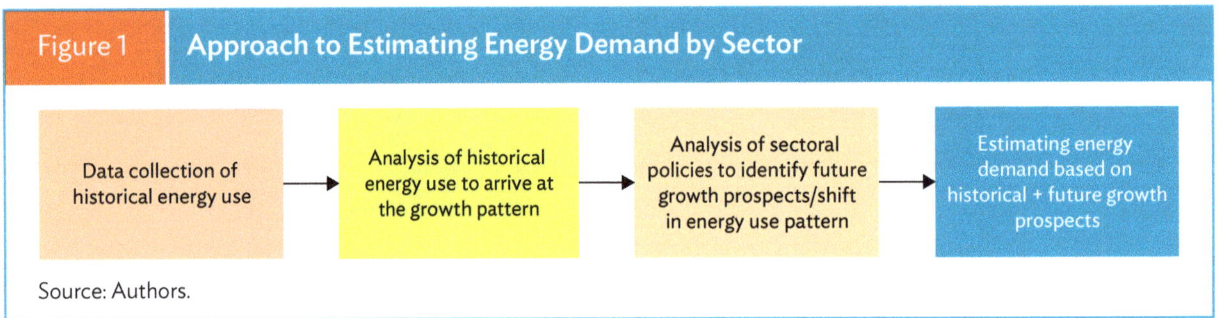

Source: Authors.

Total Final Energy Demand

The share of fuels in TFED for 1990 and 2019 is in Figure 2.

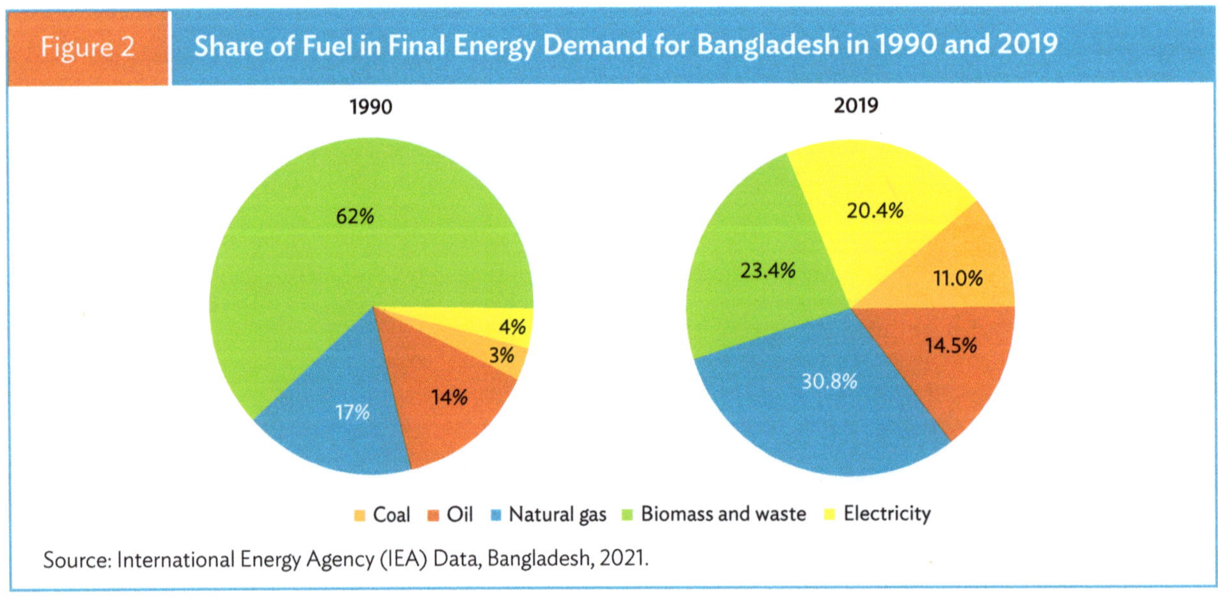

Source: International Energy Agency (IEA) Data, Bangladesh, 2021.

After 2005, there was a substantial rise in the use of natural gas, whose share has come to exceed that of biomass and waste (Figure 3).

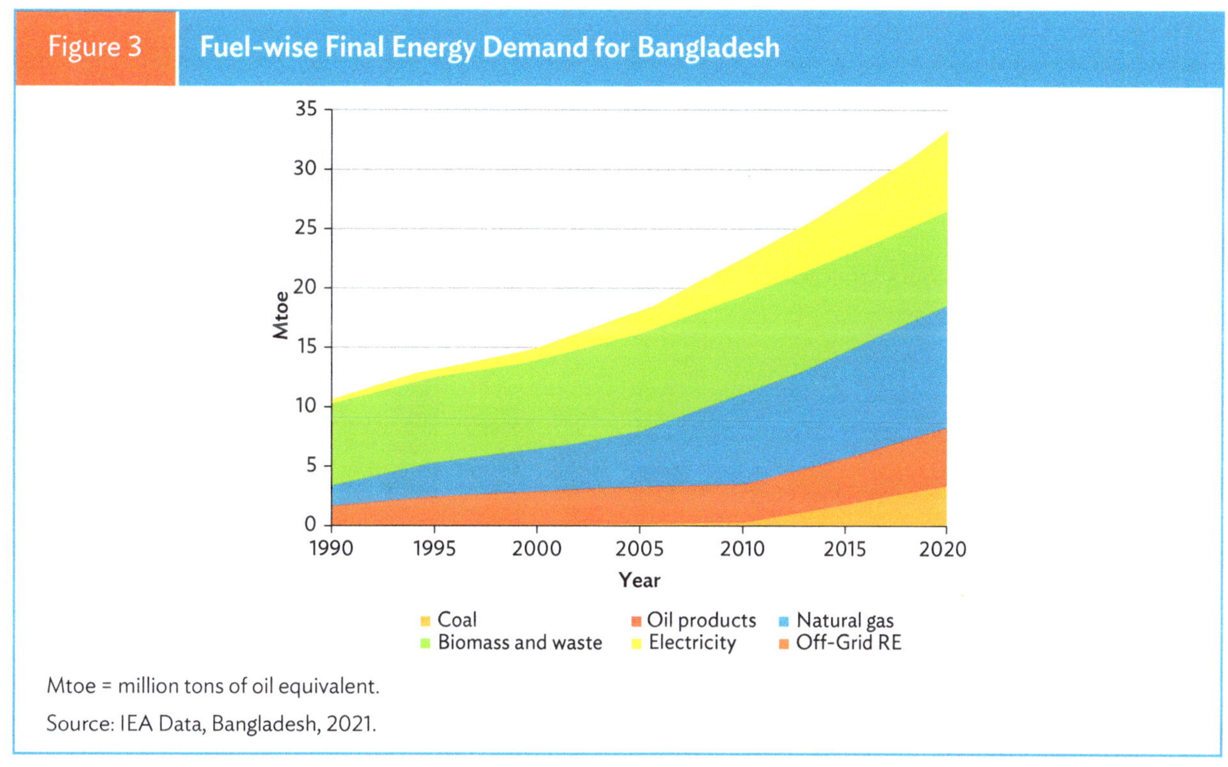

Figure 3 Fuel-wise Final Energy Demand for Bangladesh

Mtoe = million tons of oil equivalent.

Source: IEA Data, Bangladesh, 2021.

A different picture is provided by data from the Sustainable and Renewable Energy Development Authority (SREDA), which publishes annual energy balances. These have been used in this report as a reference for validation of the base year outcomes in the model. Recent trends in fuel-wise final energy demand from this source are shown in Figure 4. Final energy in 2020–2021 reached around 40.1 million tons of oil equivalent (Mtoe) from 29.9 Mtoe in 2014–2015.

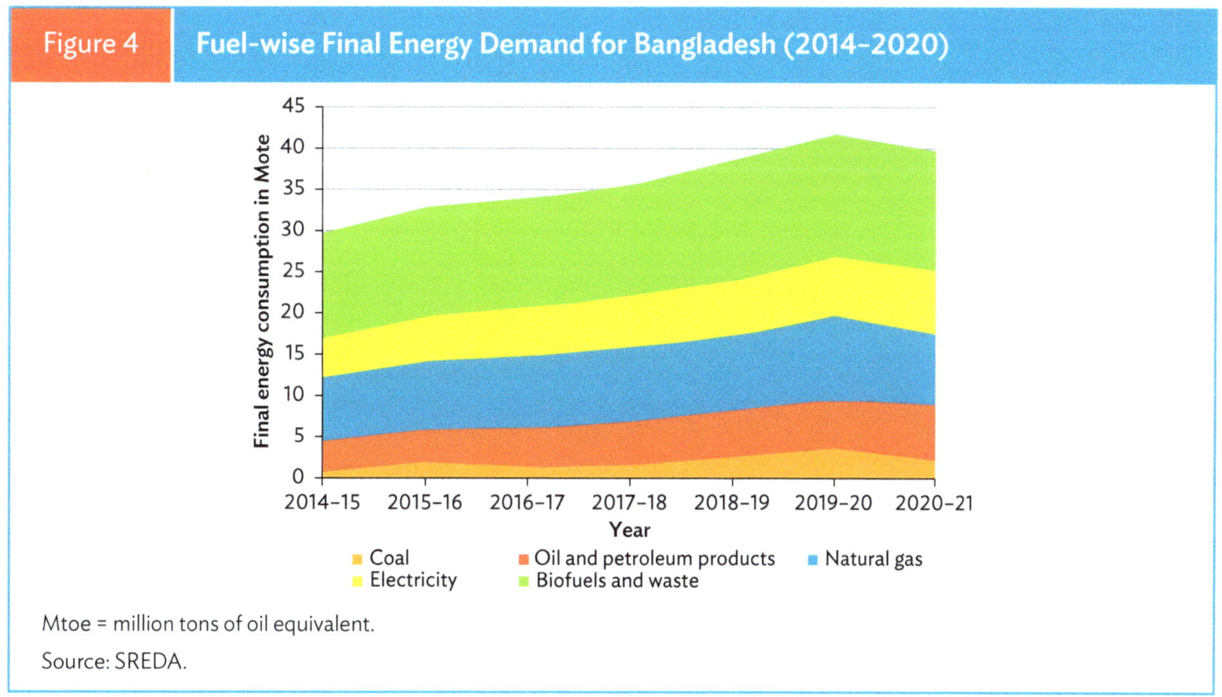

Figure 4 Fuel-wise Final Energy Demand for Bangladesh (2014–2020)

Mtoe = million tons of oil equivalent.

Source: SREDA.

Business-as-Usual Scenario

Under the BAU scenario, natural gas, electricity, and oil continue to be the primary fuels for meeting energy demand over the period until 2050, with increasing penetration of electricity and a gradual shift from conventional biomass to other fuels. TFED reaches 65.4 Mtoe in 2030 and 92.2 Mtoe in 2050, with gas providing 63% of final demand in 2050. Figure 5 shows the distribution of this demand across fuels over time.

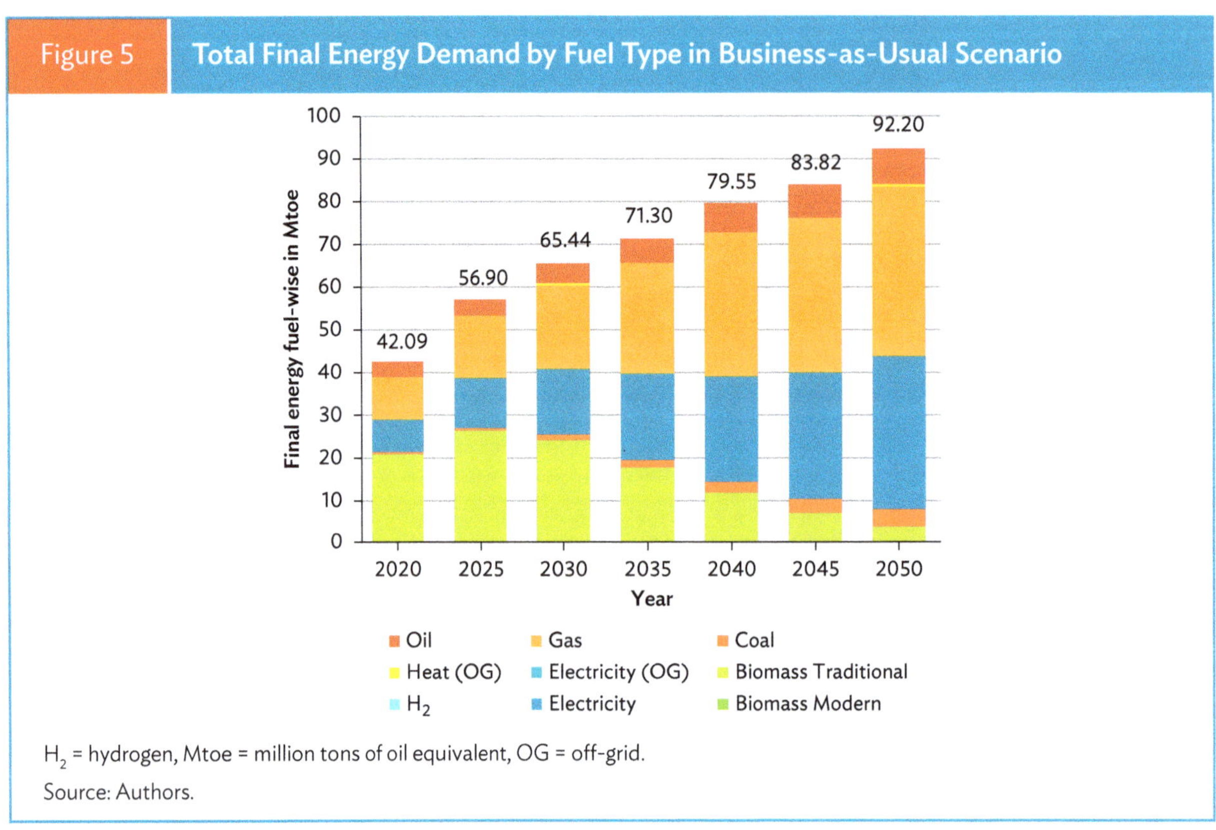

| Figure 5 | Total Final Energy Demand by Fuel Type in Business-as-Usual Scenario |

H$_2$ = hydrogen, Mtoe = million tons of oil equivalent, OG = off-grid.
Source: Authors.

Low-Carbon Scenario

In the low-carbon scenario, the TFED is lower at 55.7 Mtoe in 2030 and 87.7 Mtoe in 2050. Figure 6 shows the trend in total TFED by fuel under the low-carbon scenario.

Table 5 gives the growth of energy demand for different fuels over 2030–2050. Final energy demand for biomass is static, as part of a reduced dependence on conventional and more energy-consuming fuels. Off-grid electricity demand has a rapid growth from a low base.

Figure 7 gives the fuel mix of TFED between 2020 and 2050. While gas increases its share to 49%, that of traditional biomass drops from 20.3% in 2030 to almost zero (0.35%) in 2050, while the share of electricity increases from 28.3% in 2030 to 43.1% in 2050.

In the low-carbon scenario, by 2050, TFED is almost 5% lower than in the BAU scenario (87.8 Mtoe as compared with 92.2 Mtoe). As discussed in more detail in later sections, the key factors responsible are the increasing penetration of energy-efficient processes and appliances in the building and industry sectors, electrification, and fuel-switching to cleaner alternatives such as electric vehicles in the transport sector and the shift to electric and gas-based cooking by households. In terms of fuels, electrification as a decarbonization strategy creates a higher demand for electricity, while substitution of biomass and petroleum products by natural gas leads to higher demand for gas.

| Figure 6 | Total Final Energy Demand by Fuel Type Under Low-Carbon Scenario |

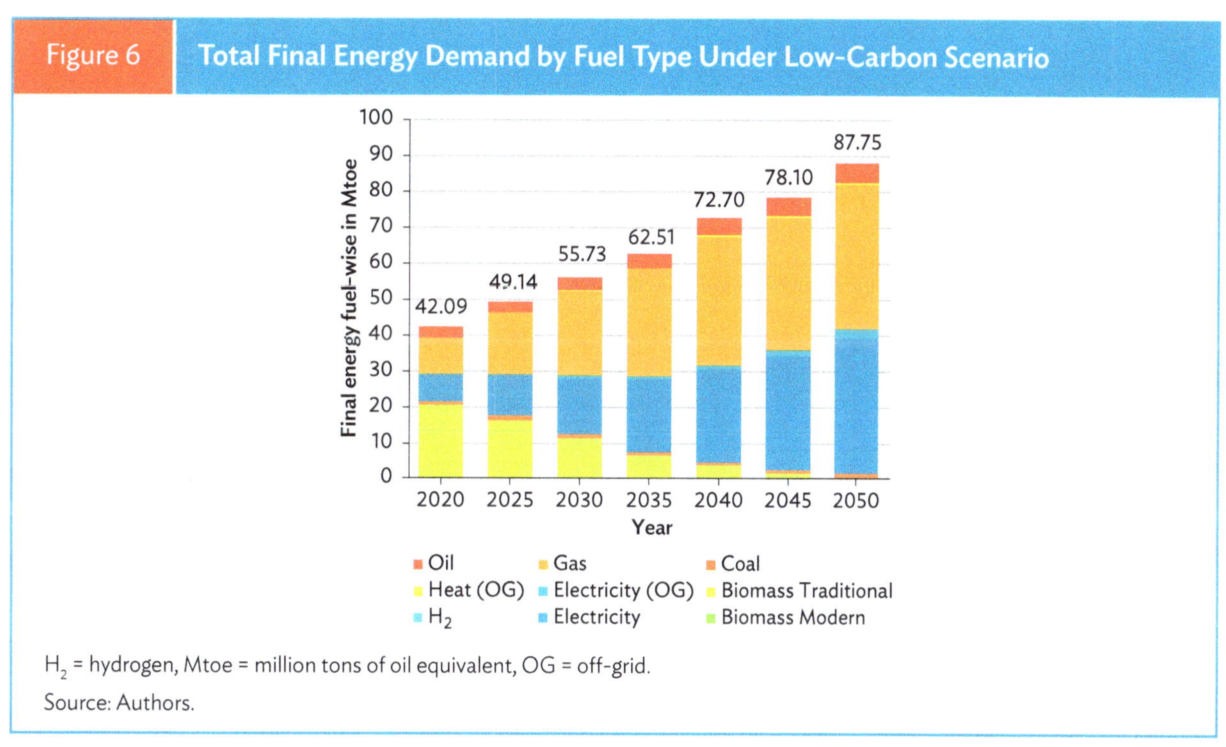

H₂ = hydrogen, Mtoe = million tons of oil equivalent, OG = off-grid.

Source: Authors.

| Table 5 | Compound Annual Growth Rate of Final Energy Demand by Major Fuels 2030–2050, Low-Carbon Scenario |

Biomass Modern	Coal	Oil	Gas	Electricity (Grid)	Electricity (Off-Grid)
0%	0.3%	2.1%	2.7%	4.5%	11.6%

Source: Compiled by authors.

| Figure 7 | Fuel-wise Share of Final Energy Demand Under Low-Carbon Scenario |

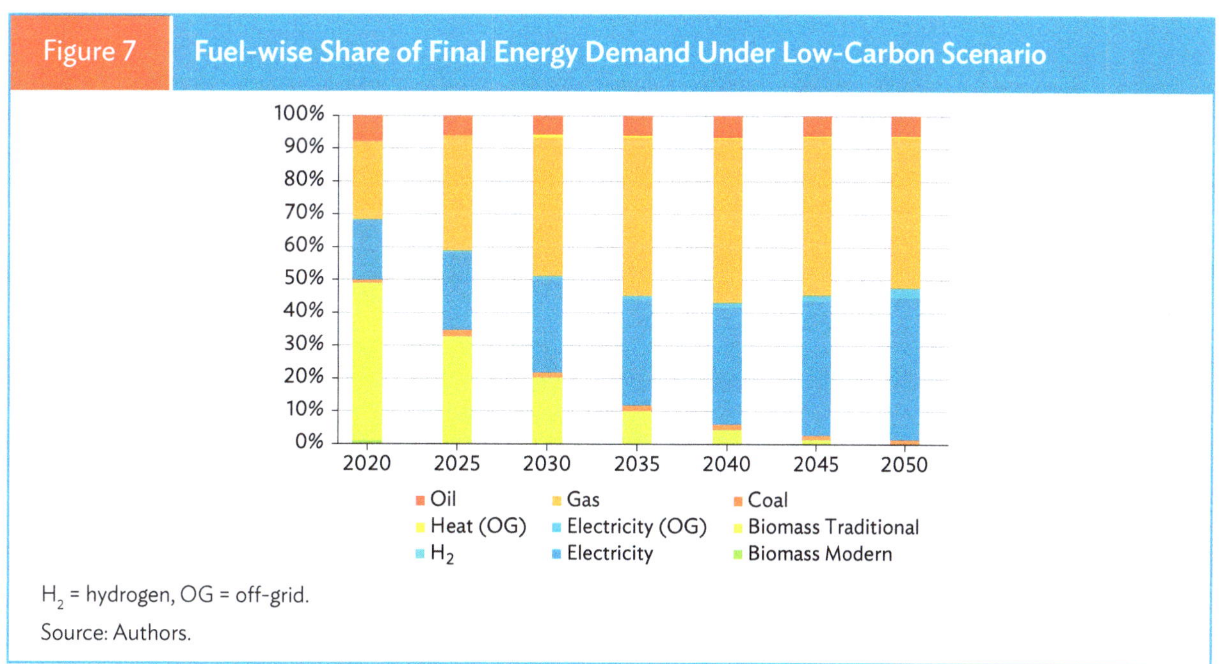

H₂ = hydrogen, OG = off-grid.

Source: Authors.

Sectoral Contribution of Final Energy Demand

Figure 8 gives the contribution to TFED of different sectors in 1990 and 2019, with the major shift being the rising share of industry over the period, reflecting industrialization and the growth of the economy.

The recent trend in sectoral final consumption according to SREDA energy balances is shown in Figure 9, with the residential sector being the by far the largest consumed, followed by industry.

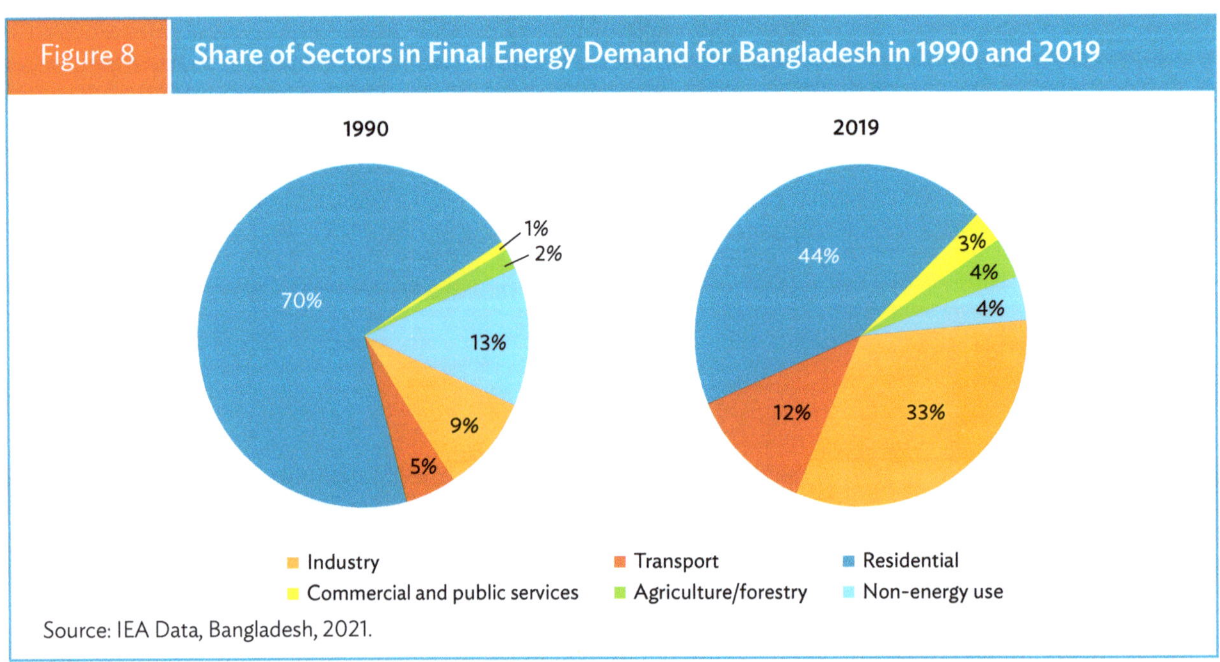

Figure 8   Share of Sectors in Final Energy Demand for Bangladesh in 1990 and 2019

Source: IEA Data, Bangladesh, 2021.

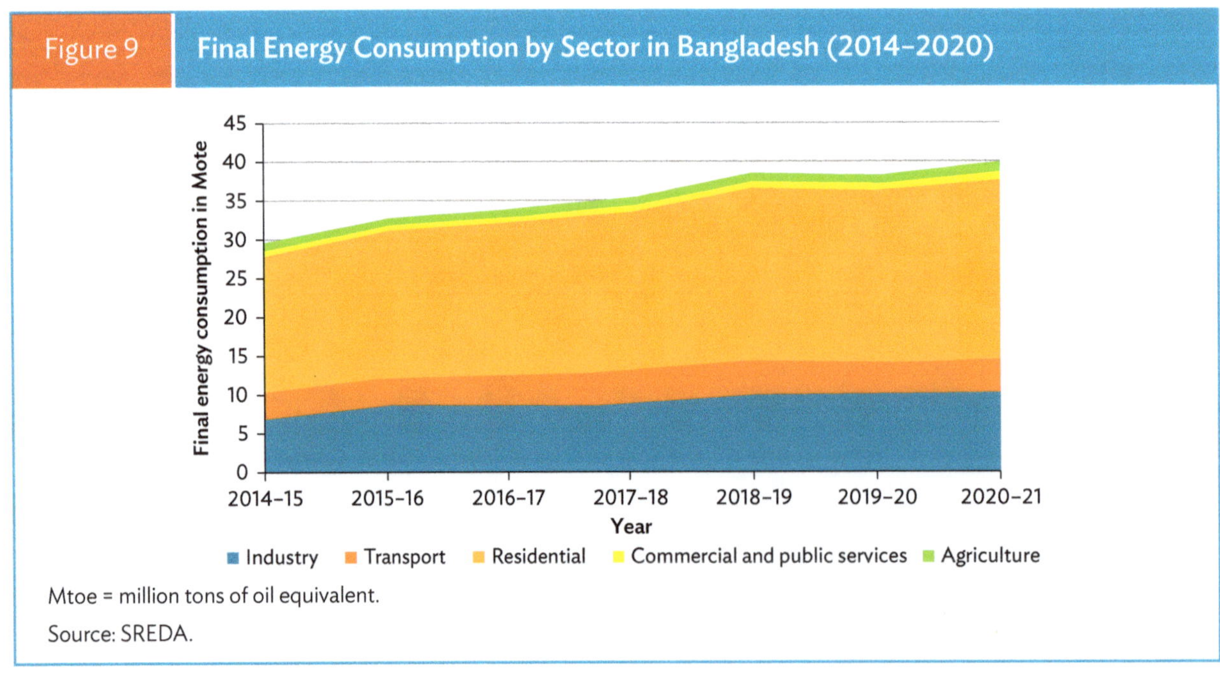

Figure 9   Final Energy Consumption by Sector in Bangladesh (2014–2020)

Mtoe = million tons of oil equivalent.
Source: SREDA.

Business-as-Usual Scenario

The model subdivides user sectors into five categories. In residential areas there is a division between cooking and non-cooking. The transport sector is categorized into two major sectors: passenger and freight. Each of these is further subdivided into four modes of transport: railways, roadways, airways, and inland waterways. For the BAU scenario growth rates of energy demand by sector between 2030 and 2050 are in Table 6, while Figure 10 gives the overall trend of energy consumption in different sectors.

The TFED is projected to grow by 1.7% annually, increasing from 65.4 Mtoe in 2030 to 92.2 Mtoe in 2050. In terms of shares, that of industry rises from 26.8% in 2030 to 36.5% in 2050 and that of the residential sector (non-cooking) also rises substantially from 10.3% in 2030 to 20.6% in 2050. The share of cooking decreases from 51% in 2030 to 28.7% in 2050 (Figure 10).

Table 6	Final Energy Sectoral Growth Rate 2030-2050 in Business-as-Usual Scenario					
	Agriculture	Commercial	Industry	Residential	Cooking	Transportation
Compound Annual Growth Rate	1.2%	2.7%	3.3%	5.4%	(1.14%)	2.9%

( ) = negative.

Source: Compiled by authors.

Figure 10	Final Energy Demand by Sector in Business-as-Usual Scenario

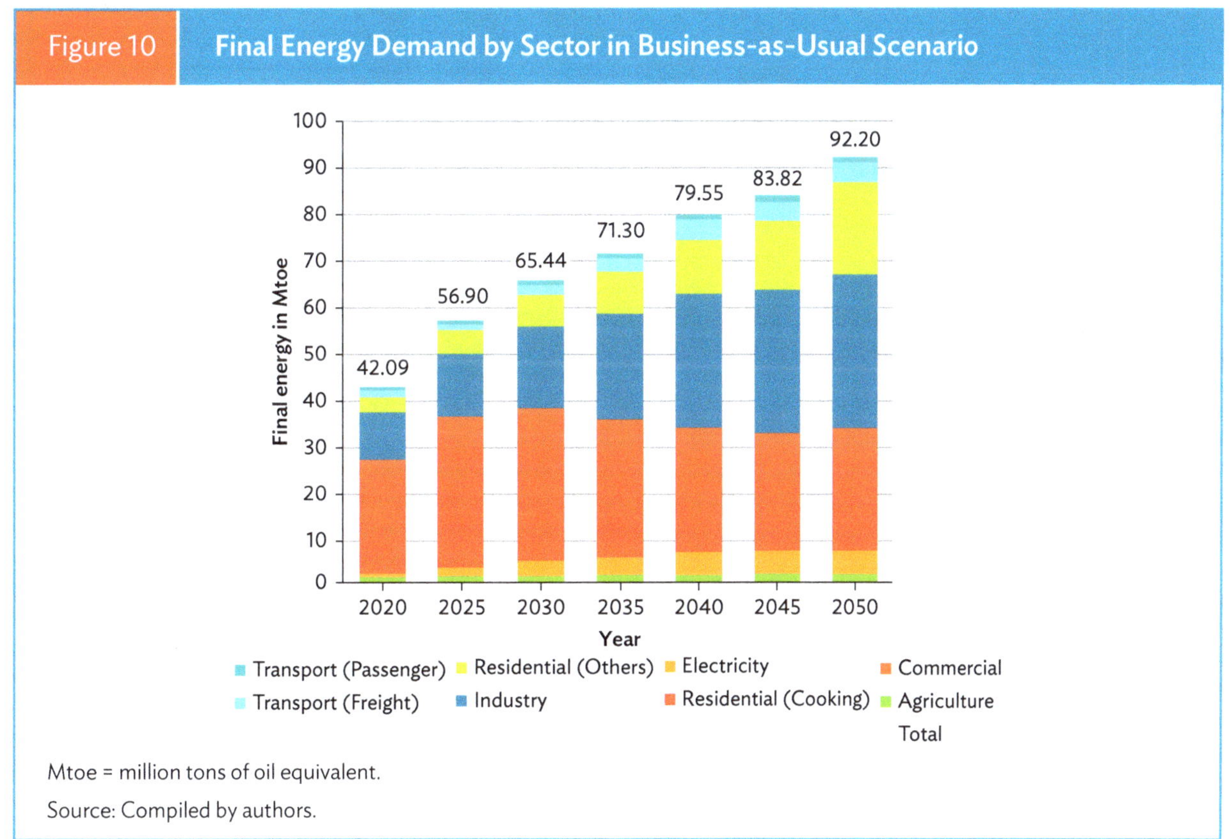

Mtoe = million tons of oil equivalent.

Source: Compiled by authors.

Low-Carbon Scenario

Under the low-carbon scenario, TFED grows from 49 Mtoe in 2025 to 88 Mtoe in 2025 (Figure 11). In terms of shares, a major shift is predicted to occur, with that of cooking falling from 42.5% in 2030 to 25.7% in 2050, due to the declining use of conventional biomass for cooking and its replacement by modern cooking appliances. Energy demand by industry increases from 17.5 Mtoe in 2030 to 32.9 Mtoe in 2050, an annual growth of 3.2%, with the corresponding share rising from 31.4% in 2030 to 37.5% in 2050 (Figure 12). Within industry, gas and electricity are the major energy commodities. Electricity demand from industry increases from 5.07 Mtoe in 2030 to 9.6 Mtoe in 2050. However, its share in final energy demand from industry remains in the range of 25–29% throughout the period. Gas is primary energy source, and its share increases from 62.7% in 2030 to 66% in 2050, while the share of coal as a fuel falls from 6% in 2030 to 3% in 2050 (Source: Compiled by authors, Figure 13). Table 7 gives compound annual growth rate (CAGR) for energy demand by sector from 2030 to 2050.

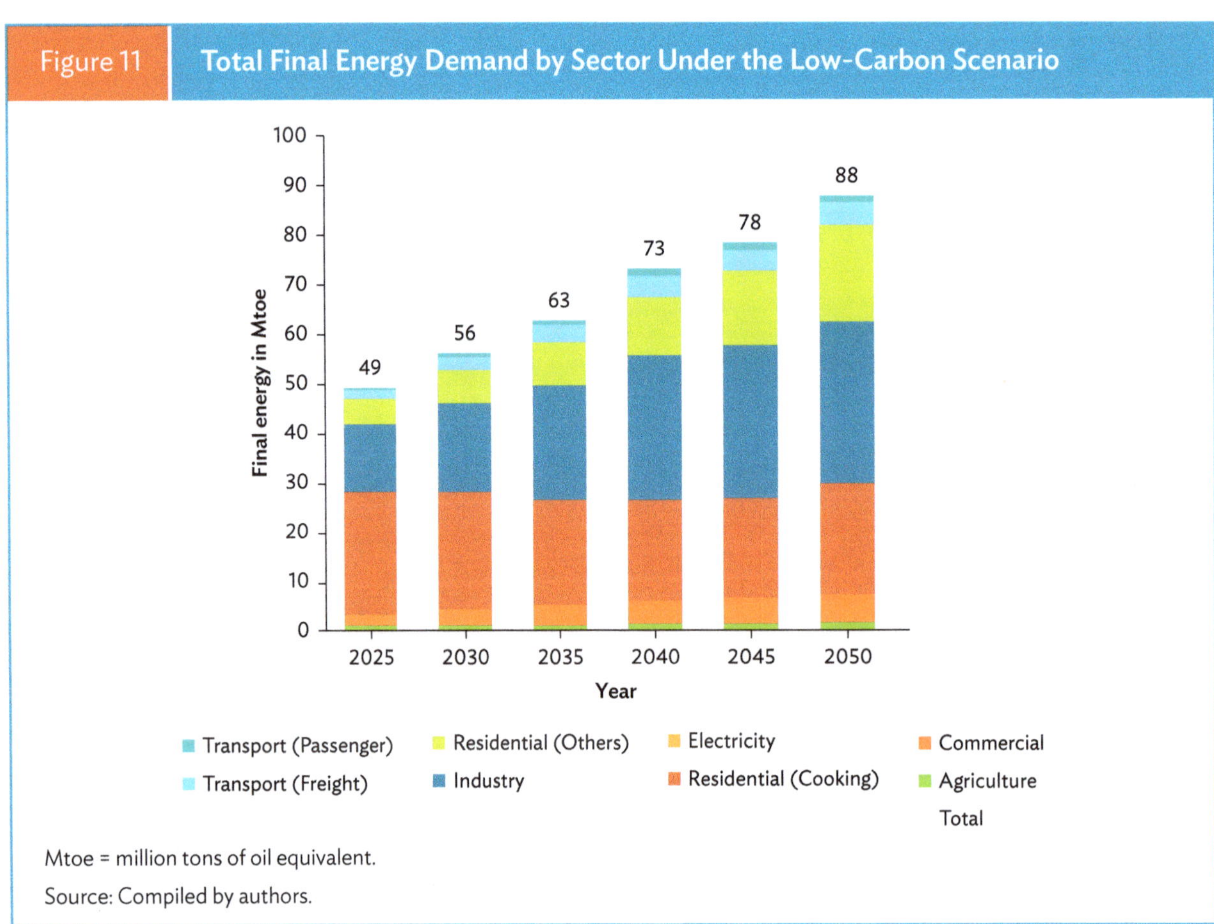

Figure 11   Total Final Energy Demand by Sector Under the Low-Carbon Scenario

Mtoe = million tons of oil equivalent.

Source: Compiled by authors.

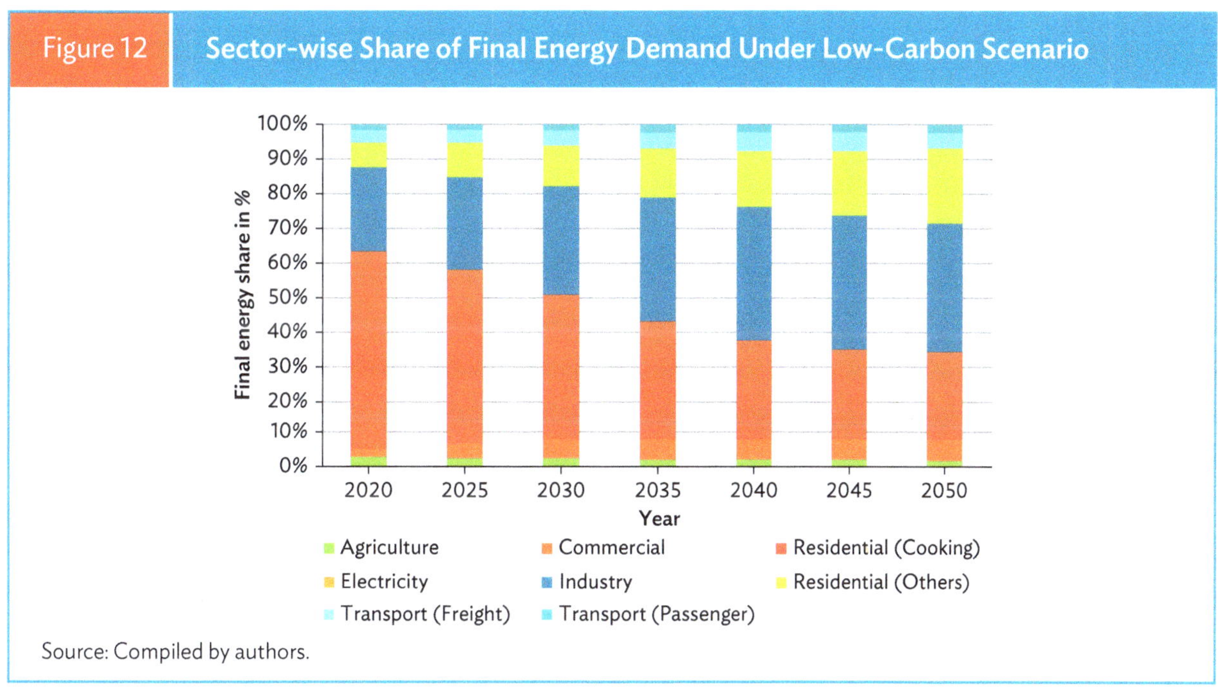

Figure 12 — Sector-wise Share of Final Energy Demand Under Low-Carbon Scenario

Source: Compiled by authors.

Figure 13 — Final Energy Demand in Industry Sector Under Low-Carbon Scenario

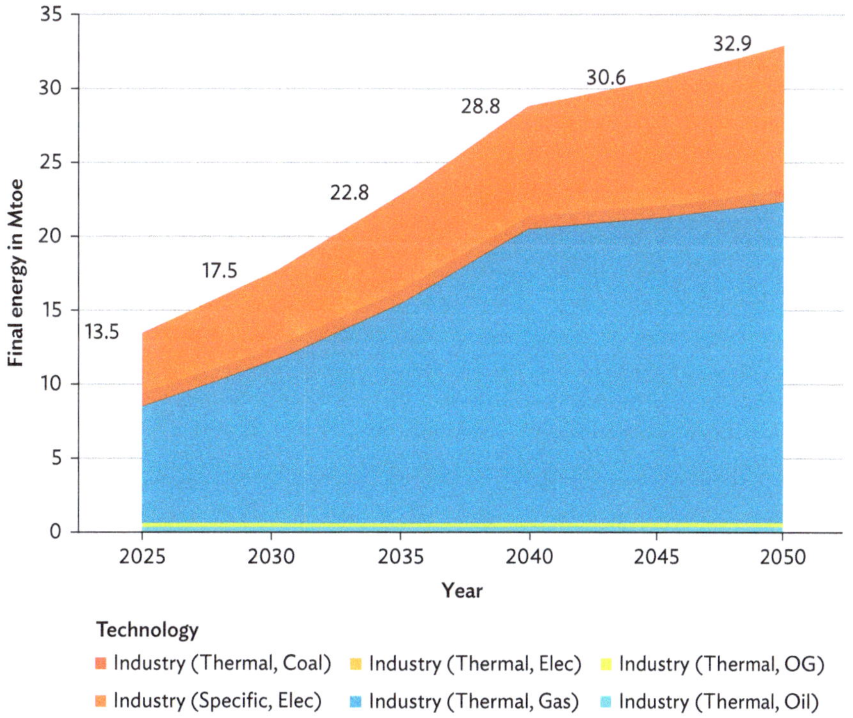

elec = electric, Mtoe = million tons of oil equivalent, OG = off-grid.

Source: Compiled by authors.

Table 7	Compound Annual Growth Rate of Final Energy Demand by Sector from 2030 to 2050				
Agriculture	Commercial	Industry	Residential (including cooking)		Transport
1.2%	2.7%	3.2%	(0.25%)		3.2%

Source: Compiled by authors.

Policy Interventions

Under the low-carbon scenario, there is a 5% reduction in final energy demand compared to the BAU scenario, as the objective of reducing emissions leads to energy savings and a transition to cleaner energy in most sectors. In agriculture, the lower final energy demand is due to the projected increased use of solar-based irrigation pump sets, with a correspondingly fall in emissions. The IDCOL, a state infrastructure investor, offers low-interest loans on solar irrigation pumps and solar plants. However, currently there is no provision for connecting the solar pumps with the grid that would generate a secondary income source for the farmers. The residential and commercial sectors are predicted to experience a decline in energy demand from buildings under the low-carbon scenario, driven by the assumption of slightly stronger implementation of building codes in residential and commercial buildings. This is over and above the progress achieved through the implementation of current policies and initiatives, like the BNBC and energy efficiency labeling programs, which are in the BAU scenario. One of the major barriers affecting progress in uptake of efficient appliances and buildings in the residential sector is the high cost. Demand-aggregation and bulk procurement models could be deployed in the case of appliances to overcome these barriers in addition to the development and enforcement of MEPS and energy efficiency labeling for all appliances.

Despite the policy interventions, like the Bangladesh Country Action Plan for Clean Cookstoves (CAP), under the National Action Plan for Clean Cooking (2020–2030), the dominant cooking energy source in Bangladesh currently is still traditional biomass. The low-carbon scenario suggests that significant energy savings from the cooking sector can be achieved by moving away from inefficient biomass-based cooking to gas and electricity-based cooking. Currently, there is limited uptake of electric cookstoves in Bangladesh due to the high cost, lack of regulatory interventions, and intermittent power supply. Possible measures to increase uptake of electric cooking are leveraging green finance, demand-aggregation and bulk procurement, establishing and enforcing country-specific MEPS and energy efficiency labeling, and ensuring an adequate and reliable power supply.

In industry, the factors driving the reduction in energy demand are increasing penetration of energy-efficient equipment, fuel-switching, and electrification. The transition from coal to natural gas leads to a reduction in energy demand through more efficient processes. Typically, high cost is a barrier to low-carbon industrial growth. Private sector investment in low-carbon technologies needs to be promoted and policy interventions like feed-in-tariffs and mandatory obligations on use of renewable energy-based technology can assist to lower emissions. In transport the introduction of electric vehicles—two-wheelers, three-wheelers, and cars—can lead to a reduction in energy demand due to higher fuel efficiency. However, a transition from oil to gas-based buses in public transport causes an overall increase in final energy consumption because of the higher fuel economy of diesel buses over their compressed natural gas (CNG) counterparts.

Total Primary Energy Supply

Natural gas is a key energy source in the country providing 59% of TPES in 2019, as compared with 29% in 1990 (Figure 14).

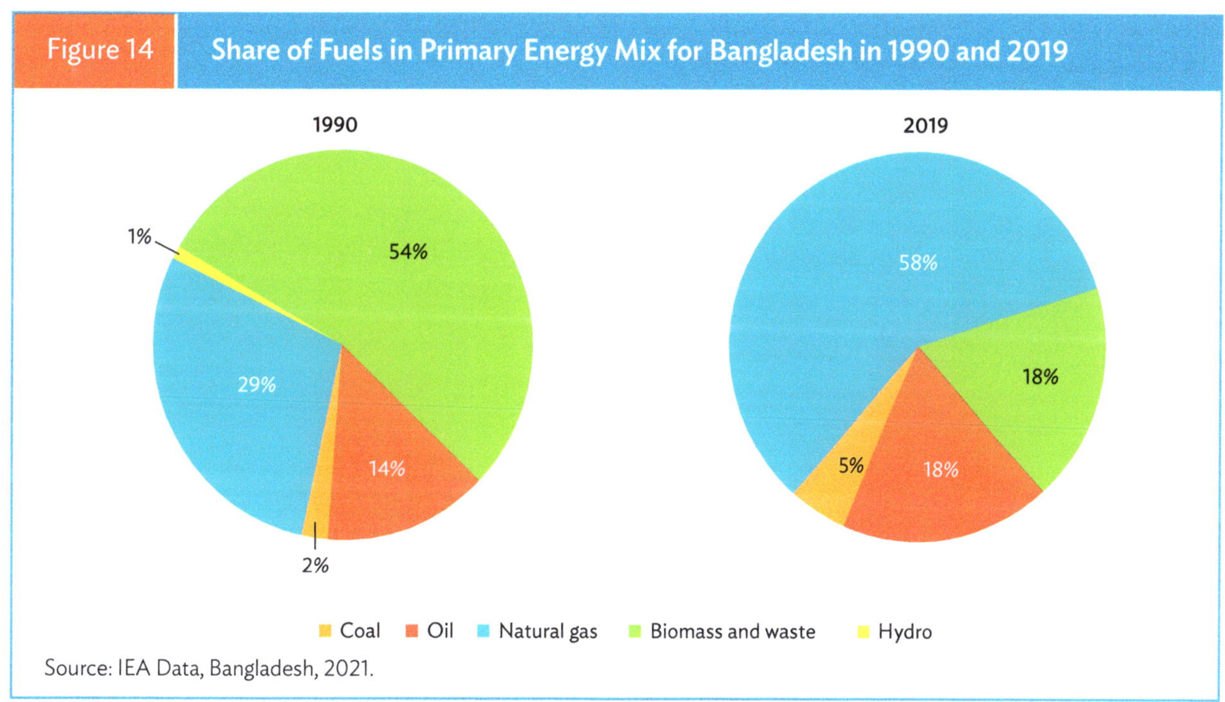

Figure 14 — Share of Fuels in Primary Energy Mix for Bangladesh in 1990 and 2019

Source: IEA Data, Bangladesh, 2021.

Figure 15 shows the trend in consumption of primary energy and fuel mix in recent few years, as published by SREDA in its annual energy balances.

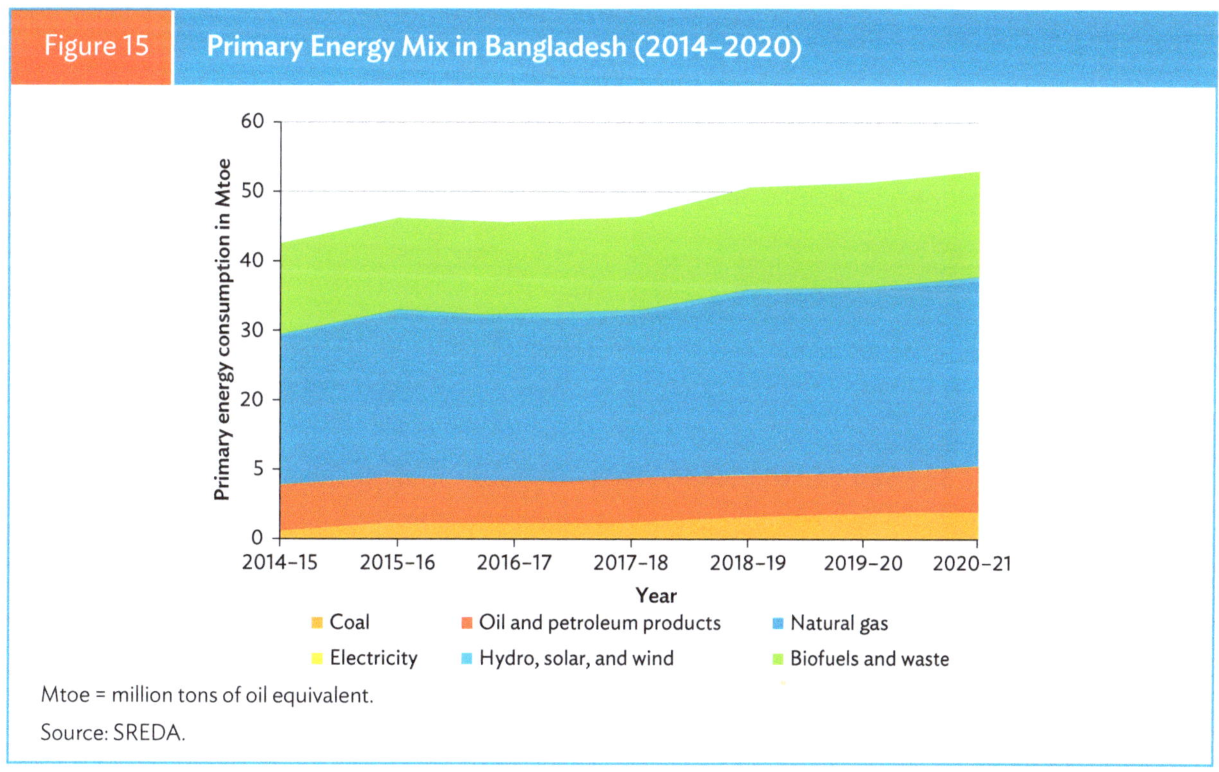

Figure 15 — Primary Energy Mix in Bangladesh (2014–2020)

Mtoe = million tons of oil equivalent.

Source: SREDA.

Business-as-Usual Scenario

Under the BAU, energy supply reaches 86.3 Mtoe in 2030 from 52.4 Mtoe in 2020 and increases to 141.2 Mtoe in 2050, growing at a CAGR of 3.4%. Figure 16 gives the projected trends in the share of domestic primary energy supply. Traditional biomass falls dramatically to 4.7% by 2050, with the share of gas rising to 63.3%.

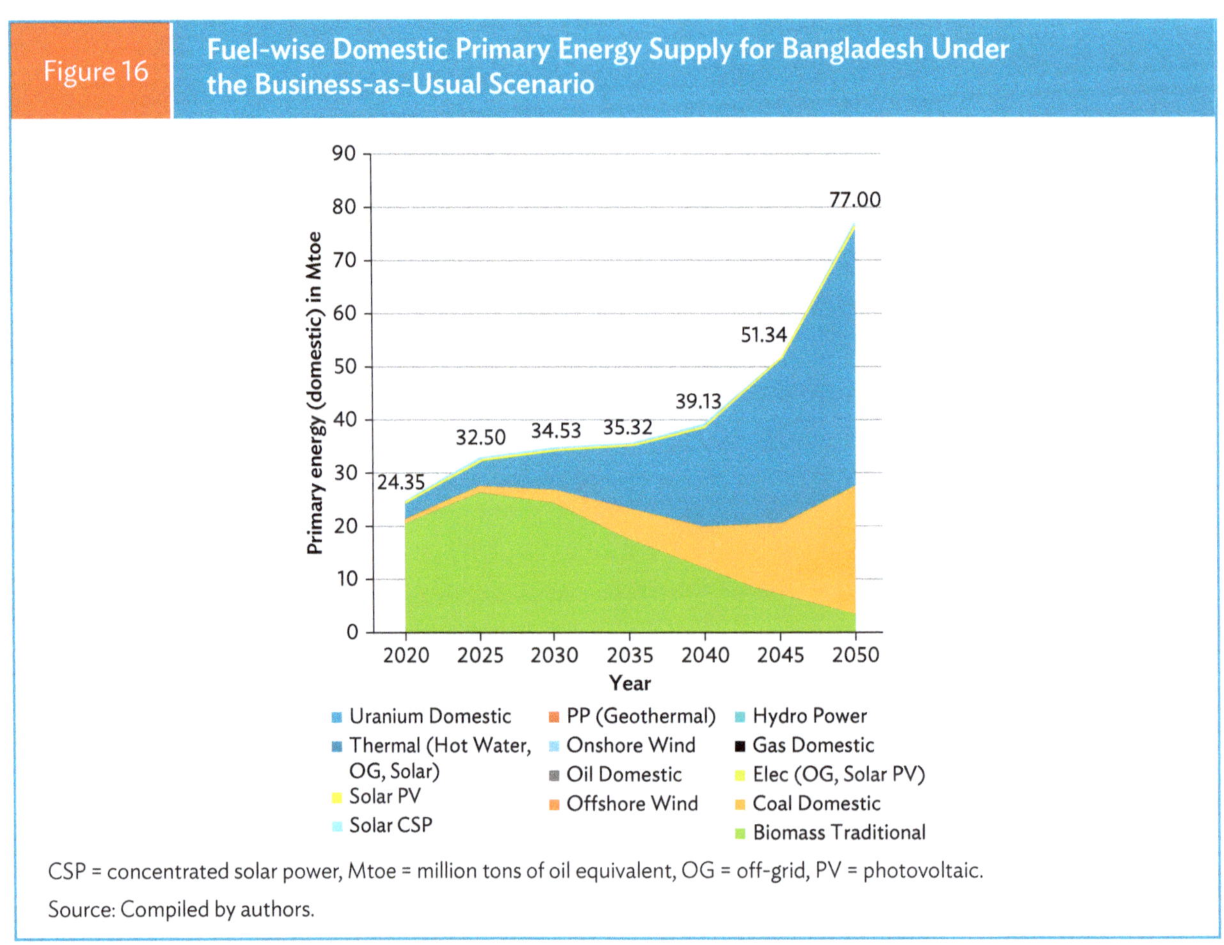

Figure 16 Fuel-wise Domestic Primary Energy Supply for Bangladesh Under the Business-as-Usual Scenario

CSP = concentrated solar power, Mtoe = million tons of oil equivalent, OG = off-grid, PV = photovoltaic.

Source: Compiled by authors.

Figure 17 shows trends in imported primary energy, which reaches 51.8 Mtoe in 2030 and 64.2 Mtoe by 2050, with a CAGR of 1%. Import dependence (imports as a share of total supply) reduces from 60% in 2030 to 45% in 2050.

Low-Carbon Scenario

Here, the TPES required to meet energy demand reaches 75.5 Mtoe in 2030 and increases to 127.8 Mtoe in 2050, a CAGR of 3% over 2030-50. Figure 18 shows the overall trends in primary energy under the low-carbon scenario. Traditional biomass still accounts for the largest share of domestic supply at around 35% in 2030, but this falls dramatically to 0.4% by 2050. Gas accounts for 35% in 2030 increasing to 76.6% by 2050. The share of coal within domestic resources decreases from 4.3% in 2030 to virtually zero (0.4%) in 2050. The total domestic primary energy reaches 31.05 Mtoe in 2030 and 79.7 Mtoe in 2050, growing at CAGR of 4.8%, with the balance between this and demand met by imports.

Figure 17 Primary Energy Import by Source for Bangladesh Under the Business-as-Usual Scenario

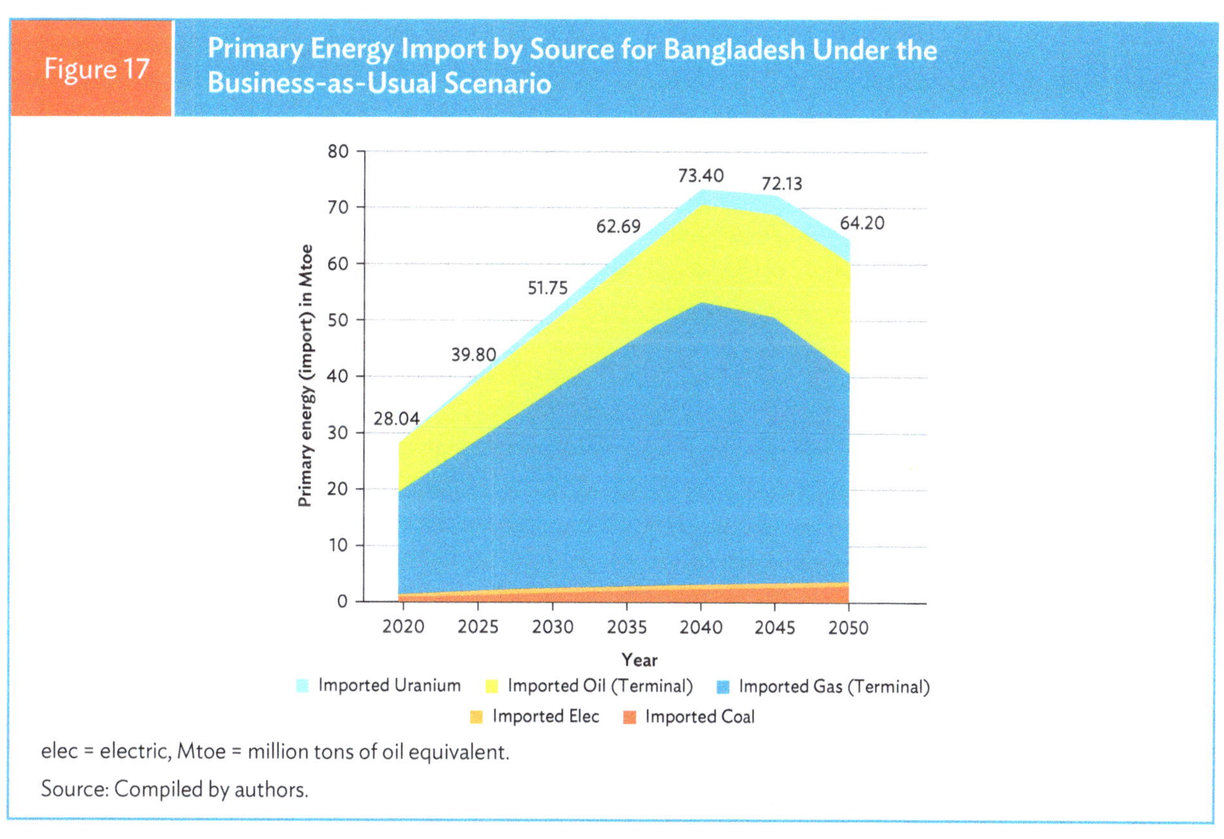

elec = electric, Mtoe = million tons of oil equivalent.

Source: Compiled by authors.

Figure 18 Fuel-wise Domestic Primary Energy Supply for Bangladesh Under Low-Carbon Scenario

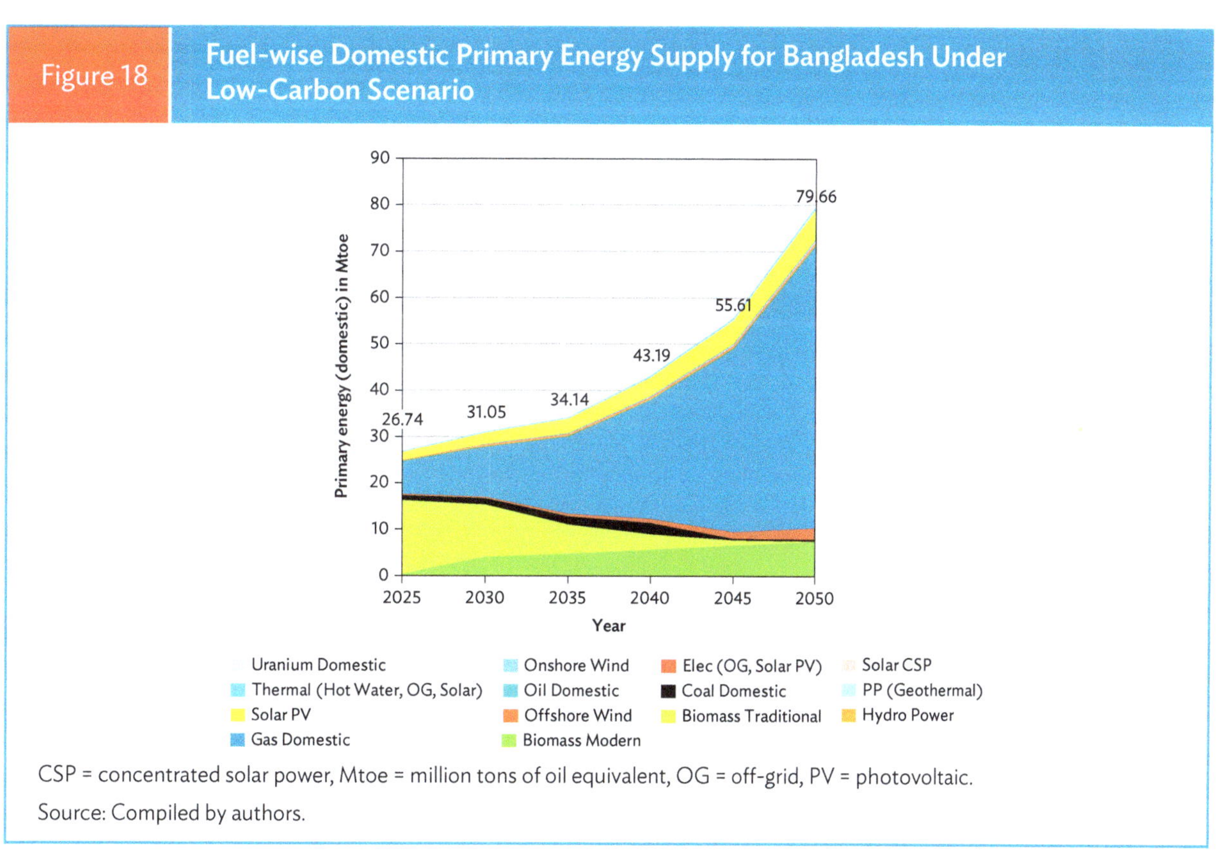

CSP = concentrated solar power, Mtoe = million tons of oil equivalent, OG = off-grid, PV = photovoltaic.

Source: Compiled by authors.

Table 8 shows the CAGR of energy supply by different domestic primary energy sources.

Table 8	Compound Annual Growth Rate of Domestic Primary Energy Supply by Various Fuel Commodities 2030–2050, Low-Carbon Scenario				
Coal	Gas	Solar PV (incl. off-grid)	Offshore Wind	Onshore Wind
(7%)	9%	6.1%	3.5%	6%

( ) = negative.

Source: Compiled by authors.

Imported primary energy is 44.5 Mtoe in 2030 and 48.1 Mtoe in 2050 (Figure 19). Correspondingly import dependence reaches 38% by 2050. Table 9 gives the growth of energy imports by fuel.

Figure 19	Primary Energy Import by Source Under Low-Carbon Scenario

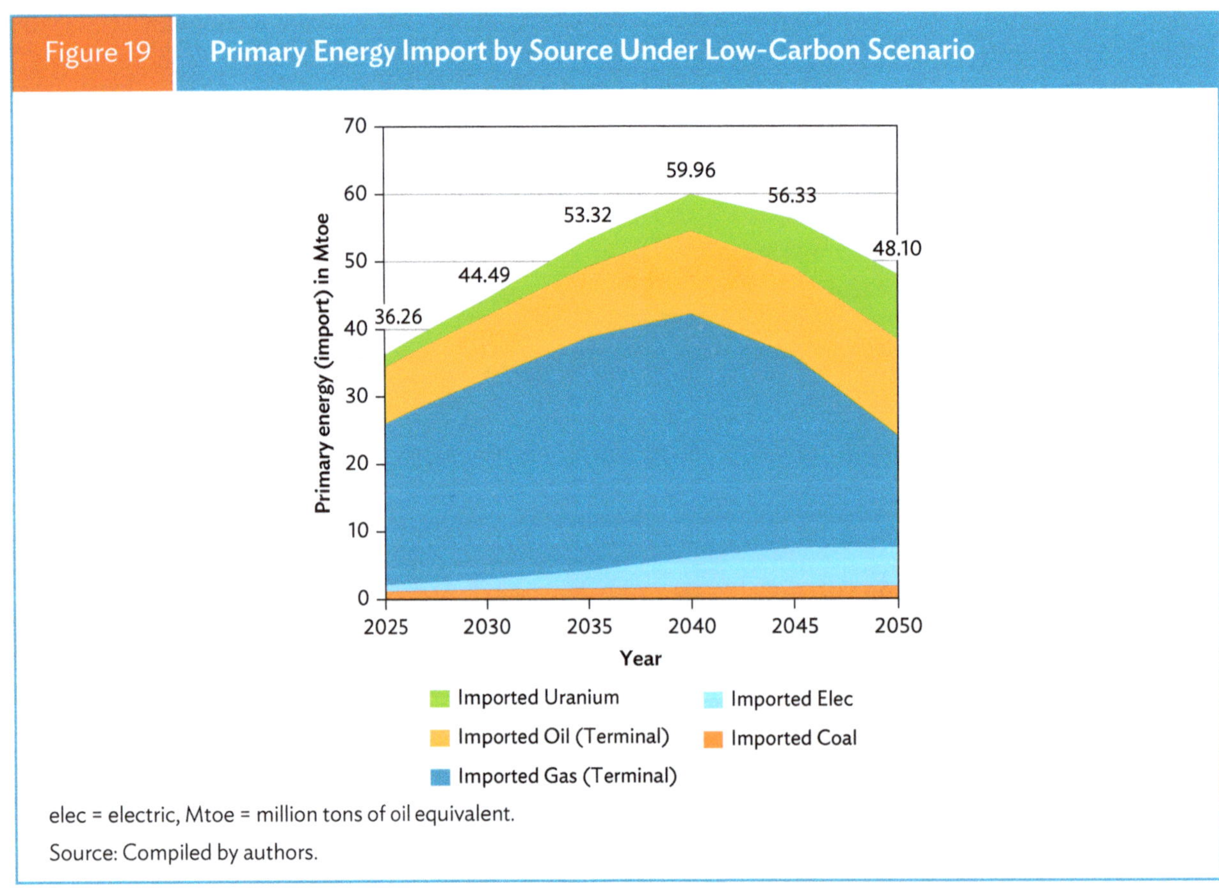

elec = electric, Mtoe = million tons of oil equivalent.

Source: Compiled by authors.

Table 9	Compound Annual Growth Rate of Energy Imports by Fuel Type 2030–2050, Low-Carbon Scenario			
Oil	Gas	Coal	Electricity
2%	(3%)	1%	7%

( ) = negative.

Source: Compiled by authors.

Oil accounts for a significant share of energy imports, with its share increasing from 21% in 2030 to 30% in 2050, while the share of gas imports first rises to over 60% in 2040 and then falls to around 35% in 2050. Nuclear-based imports increase from 2.42 Mtoe in 2030 to 9.7 Mtoe in 2050. Hence, the low-carbon scenario suggests that if the ambitious climate mitigation goals built into the model are achieved, the country will not only develop sustainably but also reduce the volume of imports from 64.2 Mtoe in 2050 under the BAU scenario to 48.1 Mtoe with natural gas becoming the major fuel import. Also, a significant push for nuclear-based power generation procured through imports is included in the low-carbon scenario to ensure energy security and clean electricity production.

Overall, within the TPES mix, covering both domestic and imported supply, in the low-carbon scenario gas is the major energy source in 2030, at 53.5%. The cumulative share of oil reduces from 12.6% in 2030 to 11.1% in 2050, while the share of coal decreases from 3.7% in 2030 to 1.7 % in 2050. By the end of the period the cumulative solar PV-based primary energy supply increases from 2.72 Mtoe in 2030 to 8.96 Mtoe and energy supply from coal is expected to reduce significantly due to the coal-to-gas transition in sectors like power and industry.

Structure of Electricity Generation

Historically electricity generation in Bangladesh has been dominated by natural gas-fired power plants, with their contribution 81% of total electricity generation in 2019, with that of oil-based generation at 17% (Figure 20). In addition to this domestic generation, the country has also imported electricity from its neighbors with imports of 6,786 gigawatt-hours (GWh) in 2019.

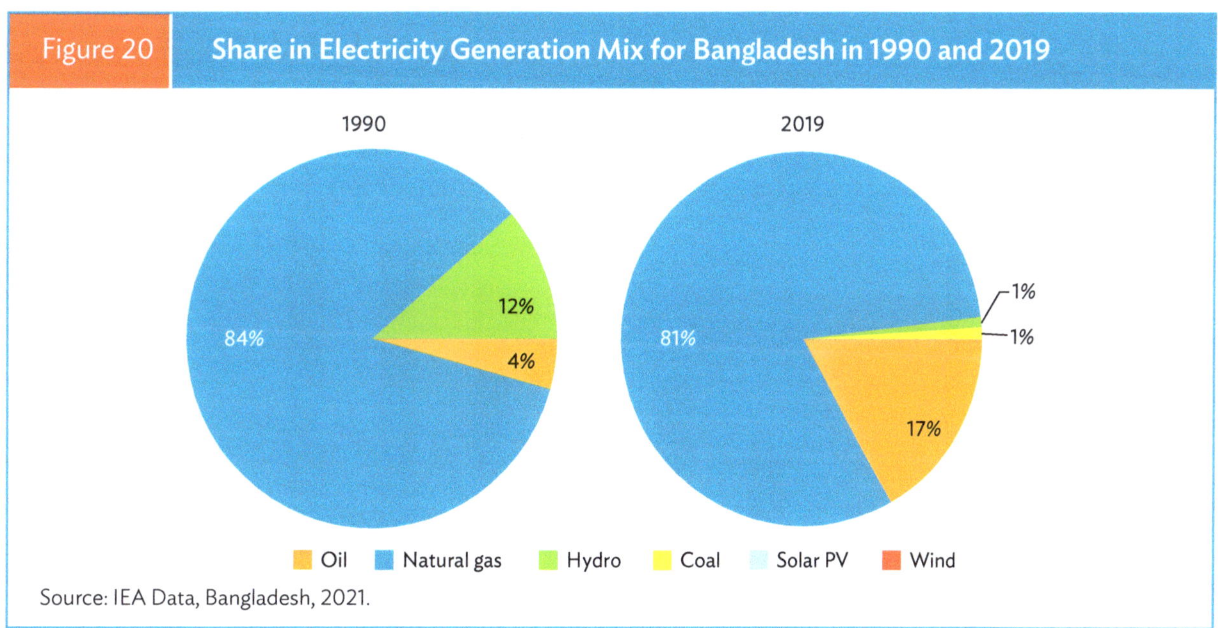

Figure 20   Share in Electricity Generation Mix for Bangladesh in 1990 and 2019

Source: IEA Data, Bangladesh, 2021.

Business-as-Usual Scenario

Under the BAU scenario, total electricity generation is projected to grow at 4.2% from 2030 to 2050, reaching 212.3 terawatt-hours (TWh) by 2030 and 484 TWh by 2050. The source-wise distribution of generation in TWh is in Figure 21, with the distribution of installed capacity in Figure 22. Coal-based power accounts for 8.3% power accounts for 70.8% of generation in 2030 and 64.2% in 2050. While the overall share of gas-based power decreases, its absolute generation increases at 3.7% annually. Oil-based power accounts for 12.2% of generation

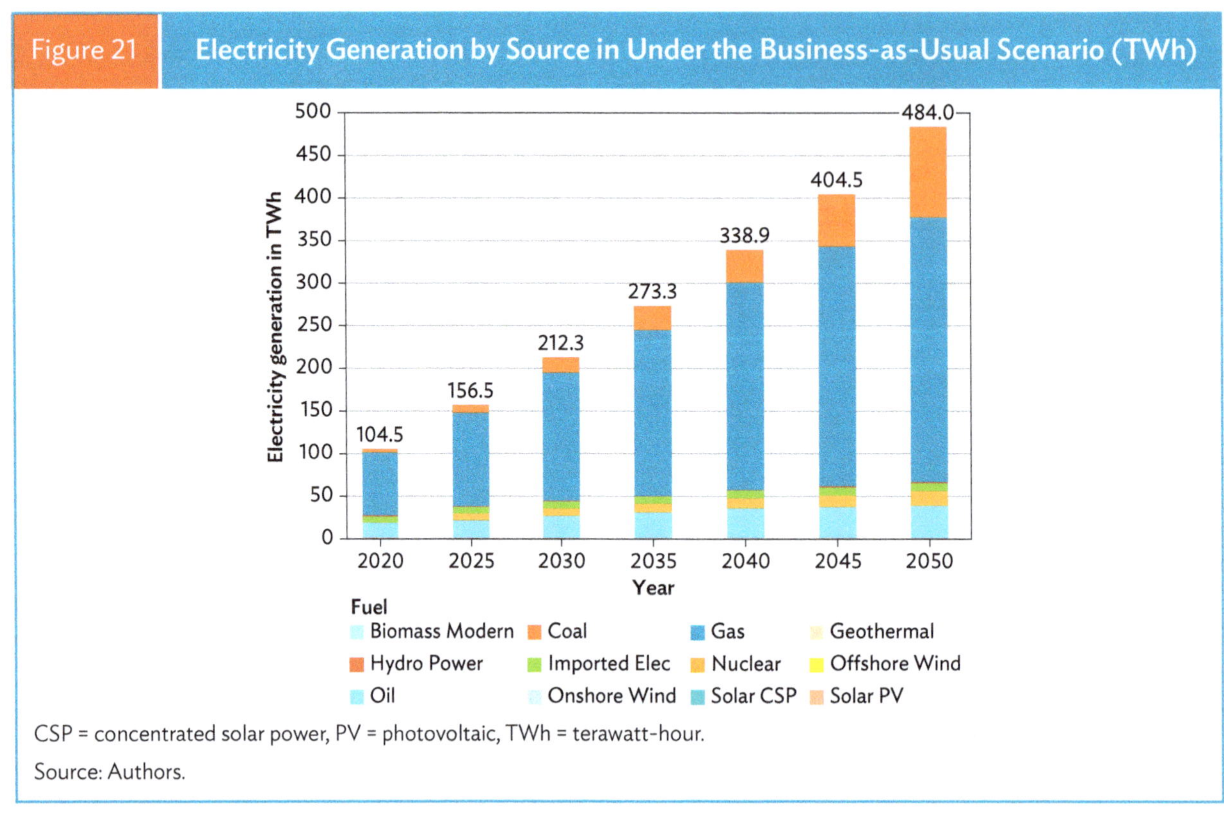

Figure 21   Electricity Generation by Source in Under the Business-as-Usual Scenario (TWh)

CSP = concentrated solar power, PV = photovoltaic, TWh = terawatt-hour.

Source: Authors.

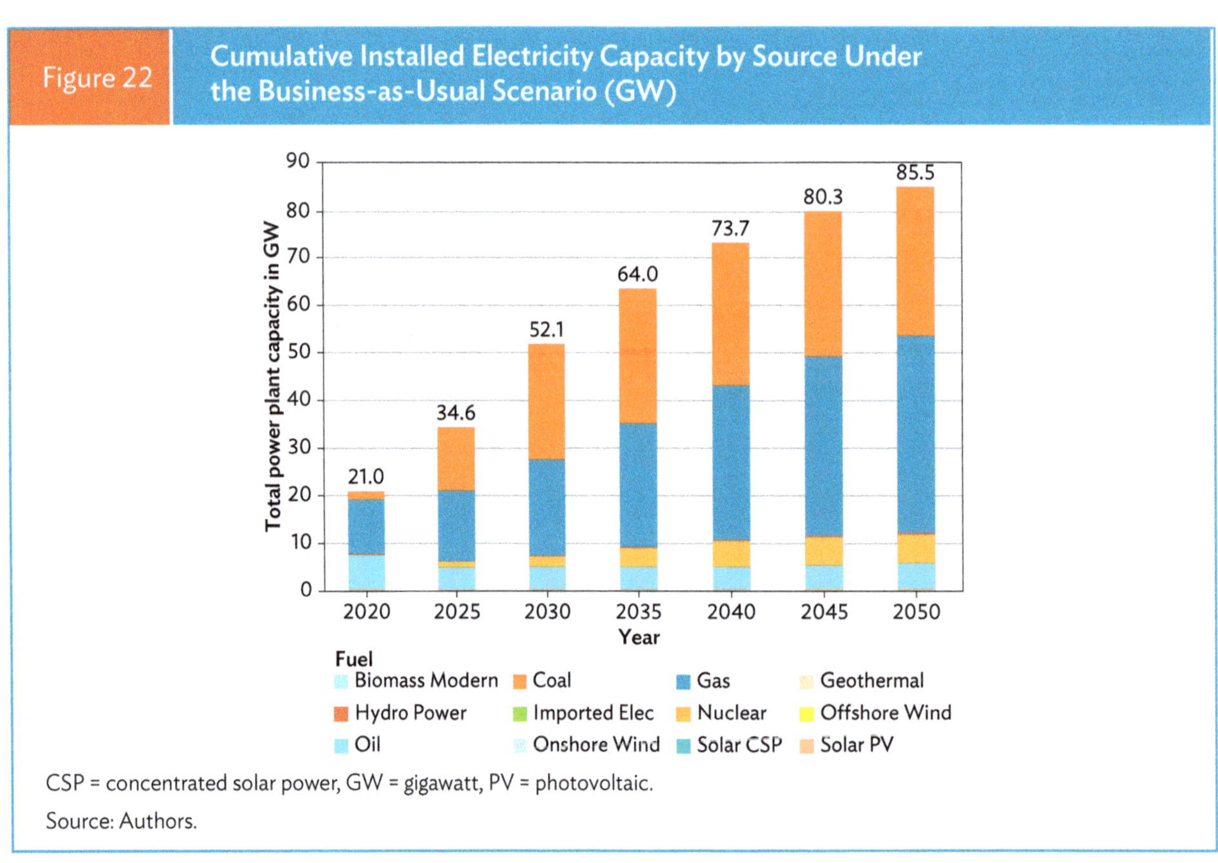

Figure 22   Cumulative Installed Electricity Capacity by Source Under the Business-as-Usual Scenario (GW)

CSP = concentrated solar power, GW = gigawatt, PV = photovoltaic.

Source: Authors.

in 2030 and 8% in 2050, used mainly to supply peak power demand. Nuclear-based generation has a modest share of 3–4% from 2025 onwards, because of new investment. Correspondingly, the generation by nuclear power plants increases from 8.53 TWh in 2030 to 17 TWh by 2050, a CAGR of 3.5%. The share of solar PV-based power generation is low (2%) under the BAU scenario.

Low-Carbon Scenario

Under the low-carbon scenario, total electricity generation reaches 214.9 TWh by 2030 and 516.1 TWh in 2050, growing at a CAGR of 4.5%, with a rising share of renewables, mainly solar and wind power. Electricity from oil-based power plants accounts for about 10% and 6% in 2030 and 2050, respectively (Figure 23). Cumulative generation from renewables increases from 49.4 TWh in 2030 to 118.6 TWh in 2050, and the share of renewables in power generation reaches 23% in 2030, of which 64% is from solar PV power, and 36% by 2050 (as opposed to only 2% in 2050 under the BAU scenario). Table 10 shows the CAGR of electricity generation from major renewable sources 2030–2050. By 2030, almost 23% of electricity generation would come from renewables, of which almost 64% is from solar PV power.

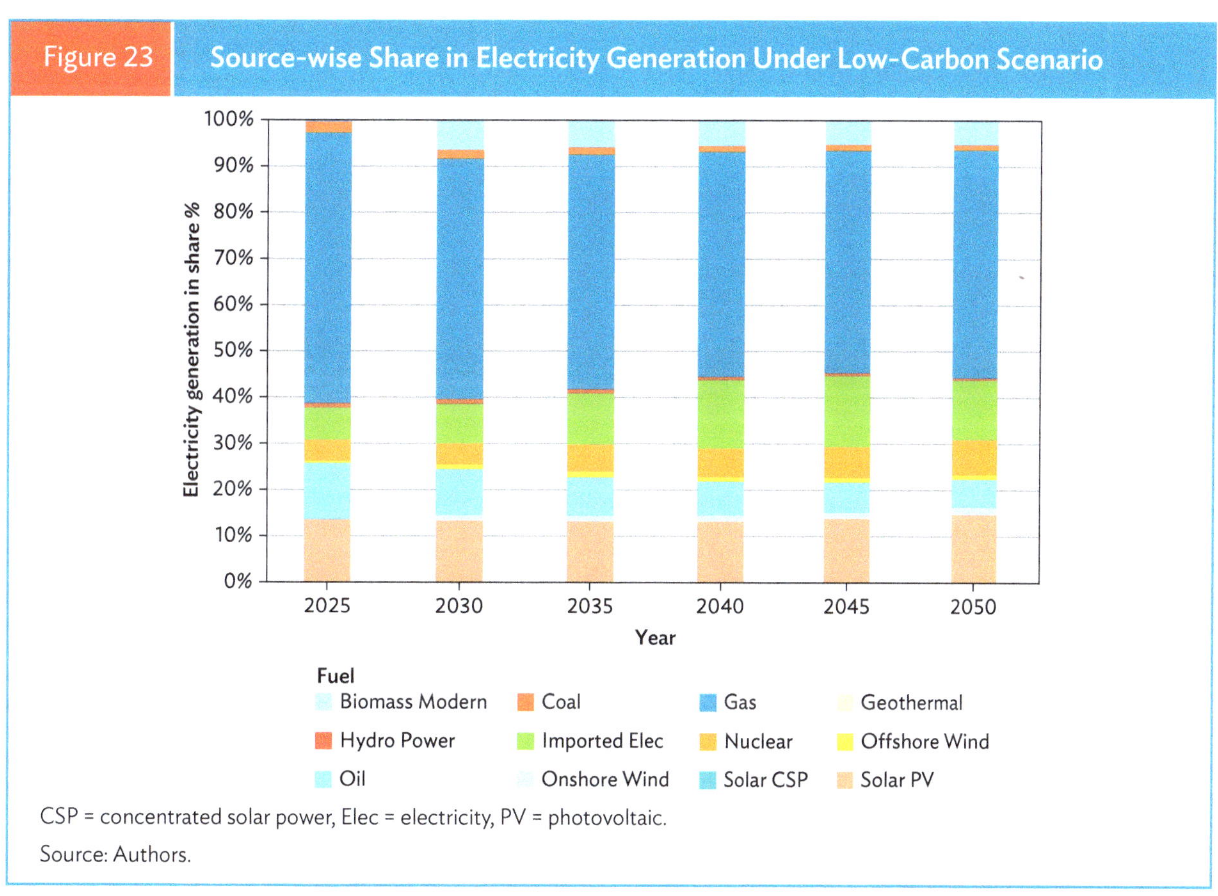

Figure 23   Source-wise Share in Electricity Generation Under Low-Carbon Scenario

CSP = concentrated solar power, Elec = electricity, PV = photovoltaic.

Source: Authors.

Table 10   Compound Annual Growth Rate of Electricity Supply from Renewable Sources 2030–2050, Low-Carbon Scenario

Solar PV	Offshore wind	Biomass	Onshore wind
5%	3.5%	3.5%	6%

Source: Compiled by authors.

The projected growth rates are based on the current level of renewable-based electricity generation and installed capacity (International Renewable Energy Agency [IRENA] 2020), as well as the targeted installed capacity of renewables (Teske et al. 2019). It is unlikely that the capacity installation of renewables could exceed the estimates projected in the low-carbon scenario (capacity installation and power generation of 64.1 GW and 118.6 TWh, respectively, by 2050) due to natural resource limitations. Therefore, regardless of the cost of renewables, the only alternative low-carbon energy source for meeting growing electricity demand, after exhausting the resource potential of renewables, is natural gas.

Figure 24	Cumulative Installed Electricity Capacity by Source Under Low-Carbon Scenario (GW)

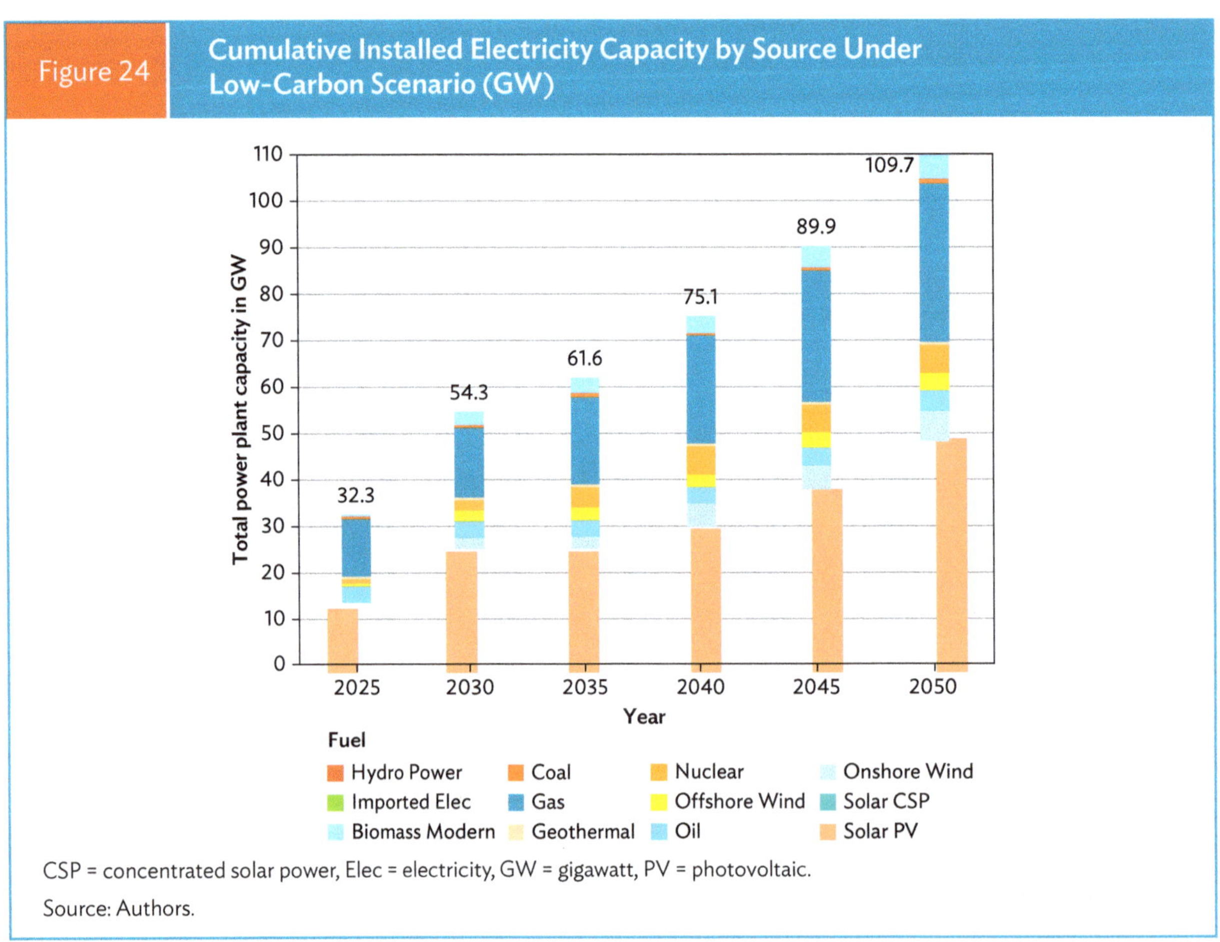

CSP = concentrated solar power, Elec = electricity, GW = gigawatt, PV = photovoltaic.

Source: Authors.

Under the low-carbon scenario, total installed capacity reaches 54.3 GW in 2030 and 109.7 GW in 2050. Figure 24 and Figure 25 show the trend in installed capacity over the period with a steady increase in capacity based on renewables, while gas remains important.

By 2030, renewables account for more than 60% of the total installed capacity, which reduces to 58.5% in 2050. The total renewable power capacity increases from about 32.6 GW in 2030 to 64.1 GW in 2050. Within renewables, solar PV accounts for the largest share of the cumulative installed capacity, reaching 25 GW in 2030 and about 48 GW by 2050. In the model the expansion of solar PV is limited to 48 GW due to the estimated resource potential of 50 GW in the country (Karim et al. 2019). In terms of wind technology, the installed capacity and electricity generation from wind offshore is projected to be lower than wind onshore, with the installed capacity of wind offshore increasing from 2.5 GW in 2030 to 3.9 GW in 2050, while that of wind onshore reaches 6.4 GW by 2050. In 2050, the share of installed gas in the cumulative mix is about 31% and oil-based power is at 4%.

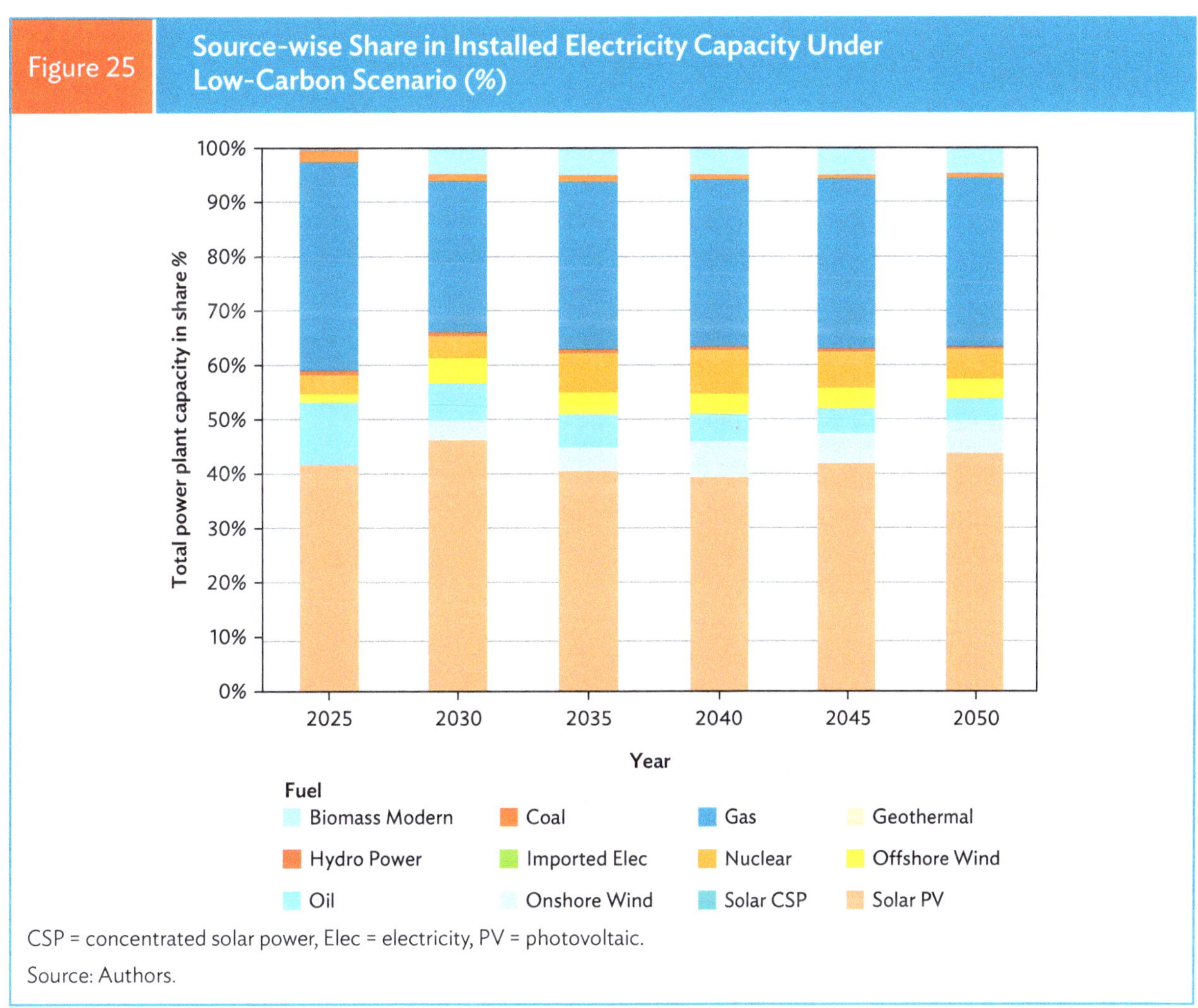

| Figure 25 | Source-wise Share in Installed Electricity Capacity Under Low-Carbon Scenario (%) |

CSP = concentrated solar power, Elec = electricity, PV = photovoltaic.

Source: Authors.

To compare the two scenarios, both electricity demand and capacity are higher in the low-carbon scenario. The total power generation capacity is estimated to be to 86 GW by 2050 under the BAU scenario, dominated by natural gas followed by coal- and oil-based power generation, along with 6 GW of nuclear power generation. In the low-carbon scenario, total power generation capacity reaches 109.7 GW in 2050 with 44% of installed capacity from solar energy. Natural gas-based capacity grows at a slower rate compared to the BAU case, while capacity based on renewables, including that based on onshore and offshore wind and biomass-based power, grows significantly.

Electrification is one of the key energy transition strategies for the country, particularly in relation to transport and residential cooking. This has implications for overall electricity demand. Under the low-carbon scenario demand reaches 516.1 TWh in 2050, a 32-TWh increase over 484 TWh in the BAU scenario. However, for electrification to be a successful strategy in meeting the government's decarbonization goals, it is essential to move away from fossil-based power generation to that based on renewables. The low-carbon scenario incorporates this shift. Under the BAU scenario, by 2050, more than 94% of the total power supply would still be derived from fossil fuels, while in the case of the low-carbon scenario, this falls to around 57%. Natural gas-based power generation continues to dominate under both scenarios; however, imports of electricity from neighboring countries increase in the low-carbon scenario to become 12.8% of the total power supply in 2050 compared with approximately 2% under the BAU scenario.

Analysis of Greenhouse Gas Emissions

Although the country has had policies to diversify its energy mix, historically GHG emissions have been rising due to the use of conventional energy sources. Emissions from energy-related sources have more than doubled since 2000, from 20.9 million tons of carbon dioxide ($MtCO_2$) to 78.3 $MtCO_2e$, in 2017, an annual growth of around 8% (Figure 26). This can be attributed to the rising energy demand created by population and industrial growth coupled with the government's vision to provide widespread electricity access at lower prices. If coal-based power generation is expanded in future this trend will only worsen.

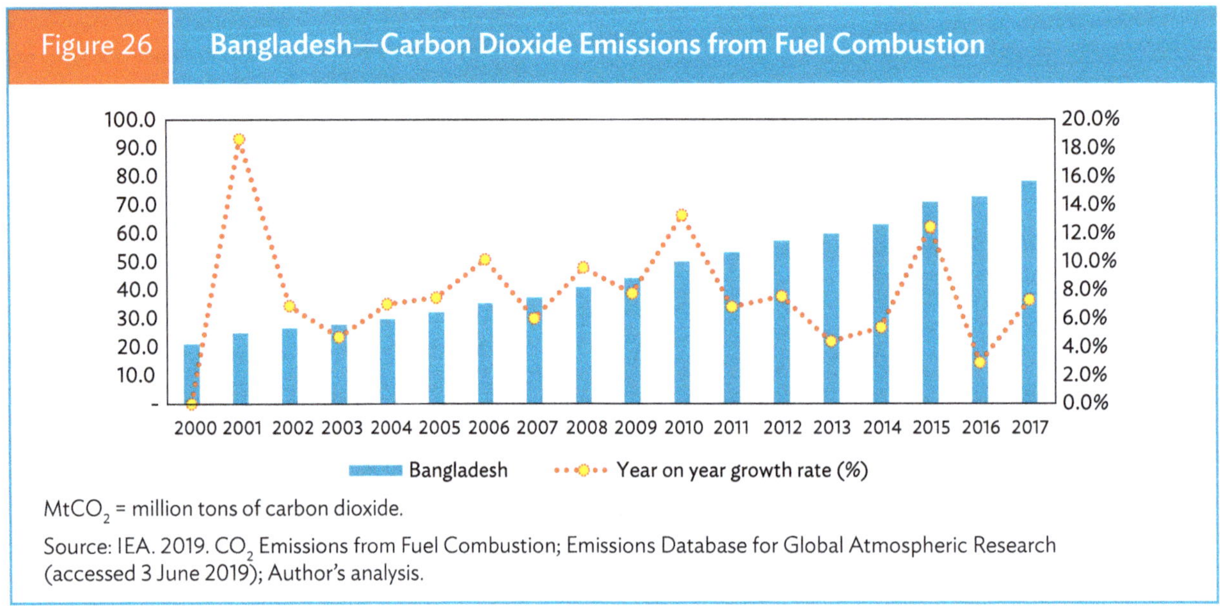

Figure 26   Bangladesh—Carbon Dioxide Emissions from Fuel Combustion

$MtCO_2$ = million tons of carbon dioxide.

Source: IEA. 2019. CO_2 Emissions from Fuel Combustion; Emissions Database for Global Atmospheric Research (accessed 3 June 2019); Author's analysis.

The country's carbon dioxide (CO_2) intensity (emissions relative to GDP) has increased from 0.12 kilograms (kg) CO_2/\$ (2010 prices) in 2007 to 0.14 kg CO_2/\$ (2010 prices) in 2017 (Figure 27).

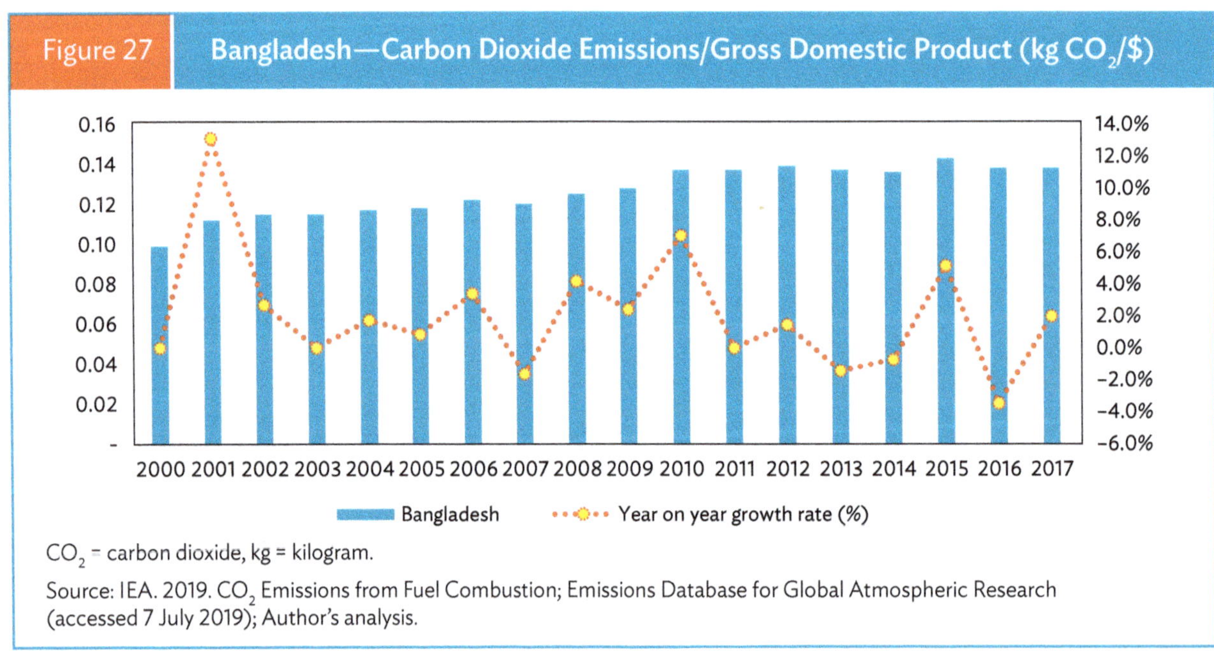

Figure 27   Bangladesh—Carbon Dioxide Emissions/Gross Domestic Product (kg CO_2/\$)

CO_2 = carbon dioxide, kg = kilogram.

Source: IEA. 2019. CO_2 Emissions from Fuel Combustion; Emissions Database for Global Atmospheric Research (accessed 7 July 2019); Author's analysis.

With rising carbon intensity, there has been a steep increase in per capita emissions, which have risen from 0.16 tons of carbon dioxide (tCO_2)/capita in 2000 to 0.48 tCO_2/capita in 2017 (Figure 28), although they remain very low compared to the global average of 4.97 tCO_2/capita/yr. (Joint Research Centre 2017).

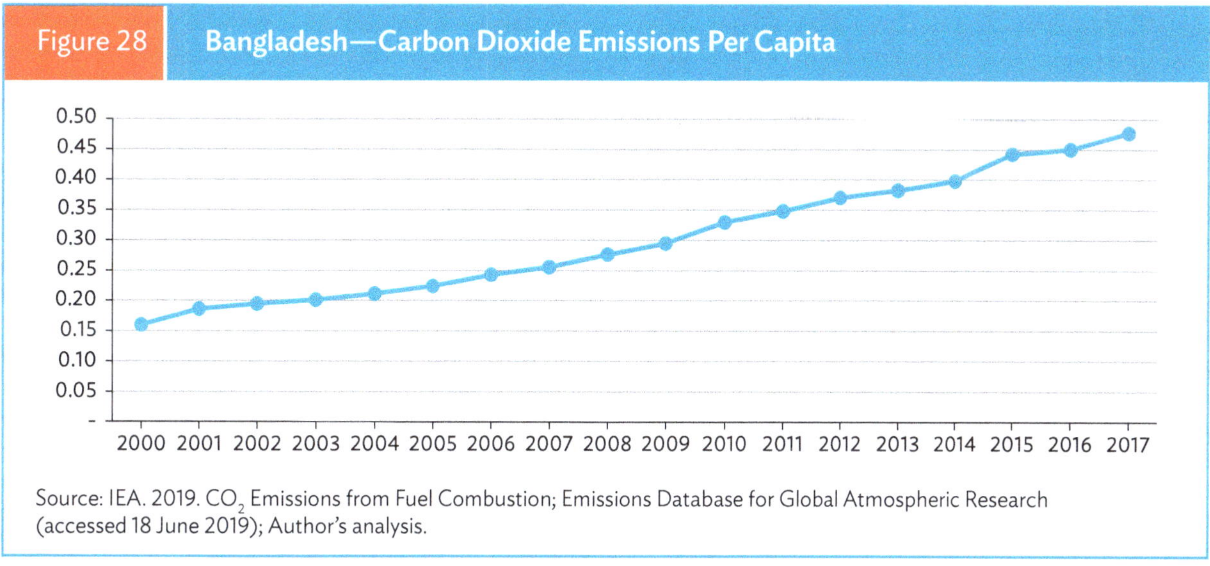

Figure 28   Bangladesh—Carbon Dioxide Emissions Per Capita

Source: IEA. 2019. CO_2 Emissions from Fuel Combustion; Emissions Database for Global Atmospheric Research (accessed 18 June 2019); Author's analysis.

At the sector level, electricity has contributed around 48% of the CO_2 emissions followed by industry, transport, and residential sectors (Figure 29) with emissions from the power increasing from 9 $MtCO_2$ in 2000 to 37 $MtCO_2$e in 2017, a CAGR of 8.7%.

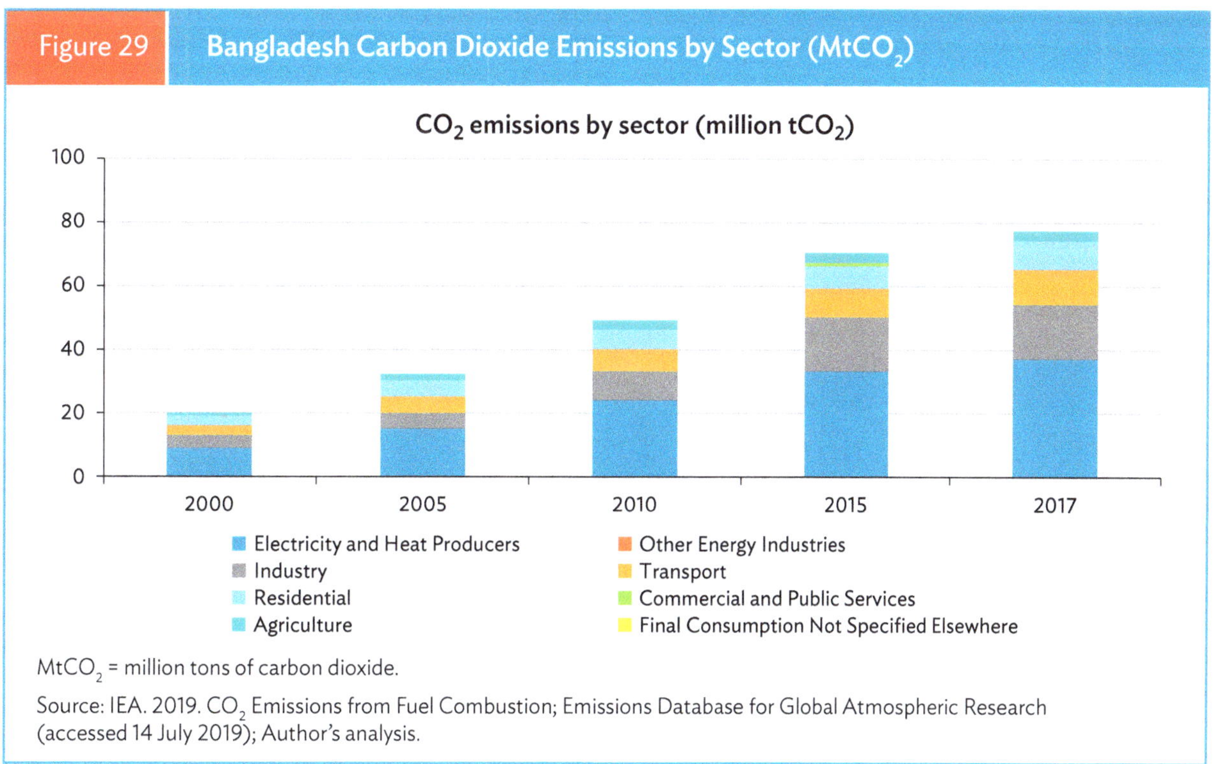

Figure 29   Bangladesh Carbon Dioxide Emissions by Sector ($MtCO_2$)

$MtCO_2$ = million tons of carbon dioxide.

Source: IEA. 2019. CO_2 Emissions from Fuel Combustion; Emissions Database for Global Atmospheric Research (accessed 14 July 2019); Author's analysis.

Business-as-Usual Scenario

Under the BAU scenario, total GHG emissions (in terms of emissions from CO_2, CH_4 (methane), and N_2O (nitrous oxide)) from the energy supply rise from 74.9 $MtCO_2e$ in 2020 to 147.5 $MtCO_2e$ in 2030 and 356.61 $MtCO_2e$ in 2050. The total GHG emissions from the supply-side increase from 79.7 $MtCO_2e$ in 2030 to 134 $MtCO_2$ in 2040 and 217.4 $MtCO_2e$ in 2050 (Figure 30).

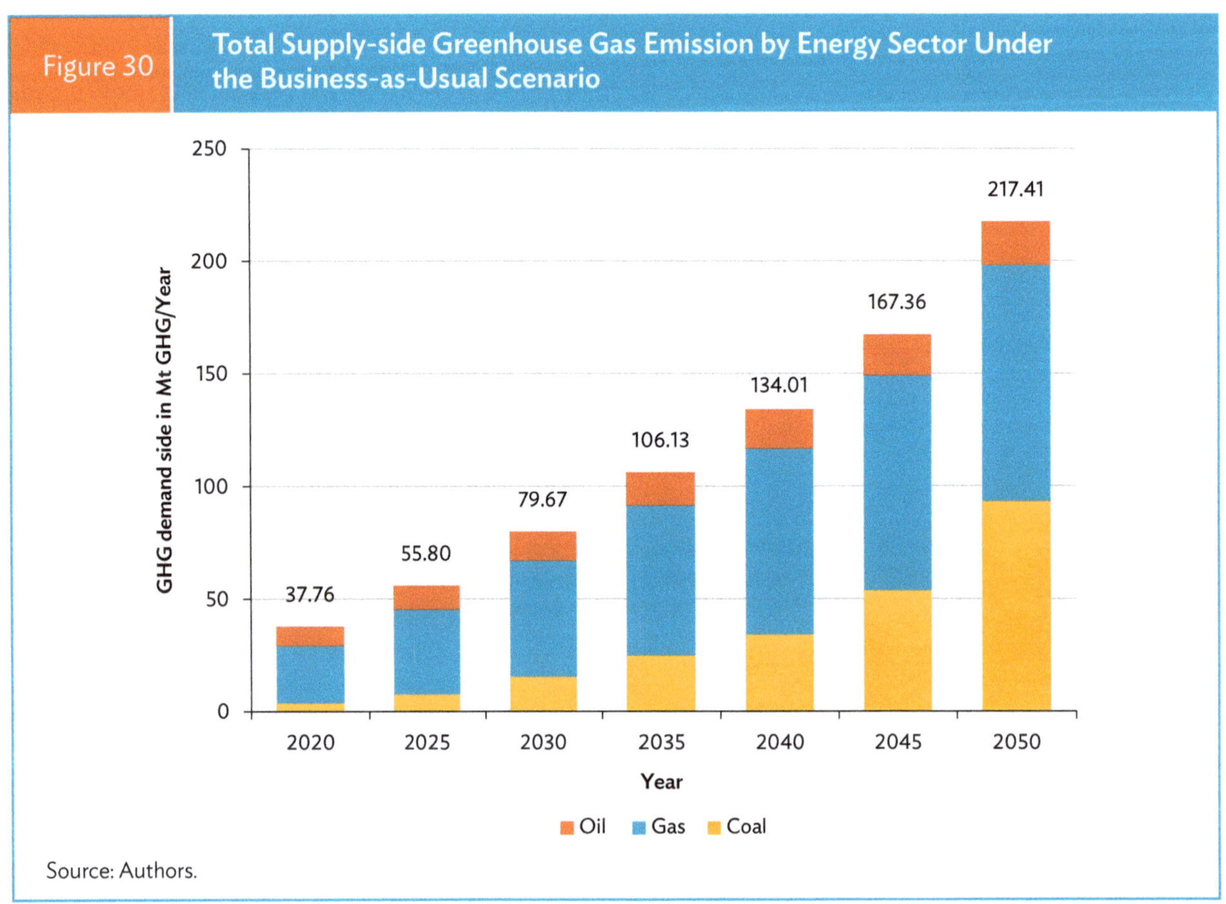

Figure 30 — Total Supply-side Greenhouse Gas Emission by Energy Sector Under the Business-as-Usual Scenario

Source: Authors.

Total annual GHG emissions from electricity generation are 74.7 $MtCO_2$ in 2030, 126.8 $MtCO_2e$ in 2040, and 209.4 $MtCO_2e$ in 2050. Of the total emissions, coal power plants emit 19% in 2030, reaching 43% in 2050. The emissions from gas power plants account for 65% in 2030 and 48% in 2050 (Figure 31).

On the demand side, in total, the demand sectors emit 67.8 million tons of carbon dioxide equivalent ($MtCO_2e$) in 2030, 112.2 $MtCO_2e$ in 2040, and 139.2 $MtCO_2e$ in 2050. The industry sector is the largest emitter and contributes 48% of emissions on the demand side in 2030 and a similar share in 2050. The share of emissions from residential cooking increase from about 31% in 2030 to about 37% in 2050, while transport's share decreases from about 16% in 2030 to about 13% in 2050. Agriculture has a low and declining share, from 4% in 2030 to 2% in 2050 (Figure 32 and Figure 33).

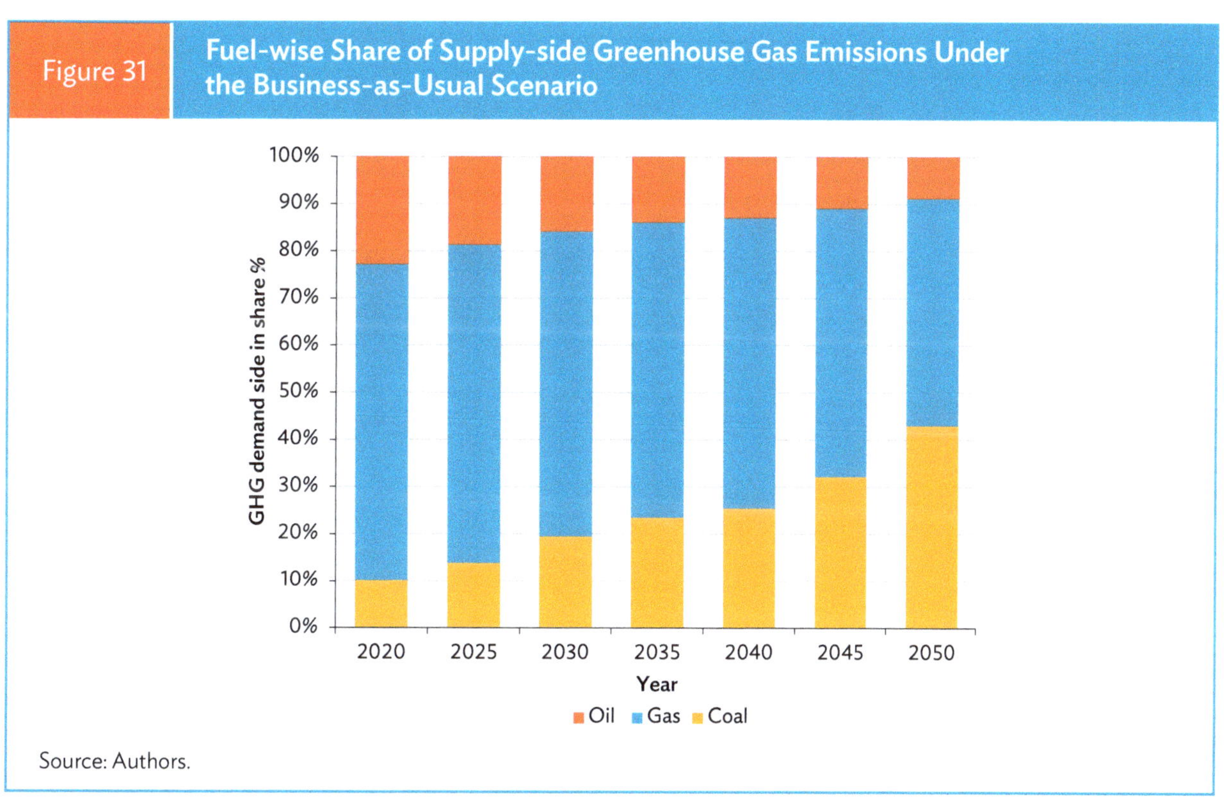

Figure 31 Fuel-wise Share of Supply-side Greenhouse Gas Emissions Under the Business-as-Usual Scenario

Source: Authors.

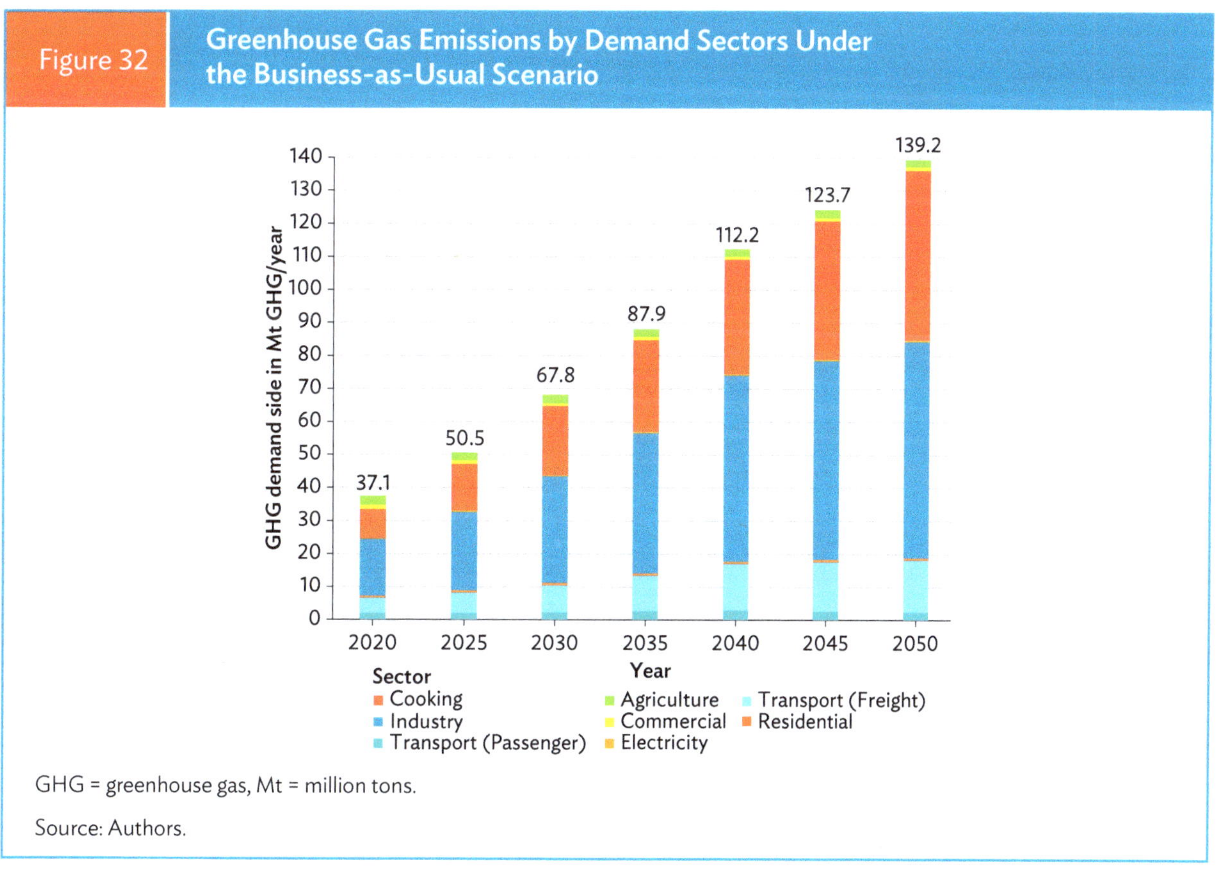

Figure 32 Greenhouse Gas Emissions by Demand Sectors Under the Business-as-Usual Scenario

GHG = greenhouse gas, Mt = million tons.

Source: Authors.

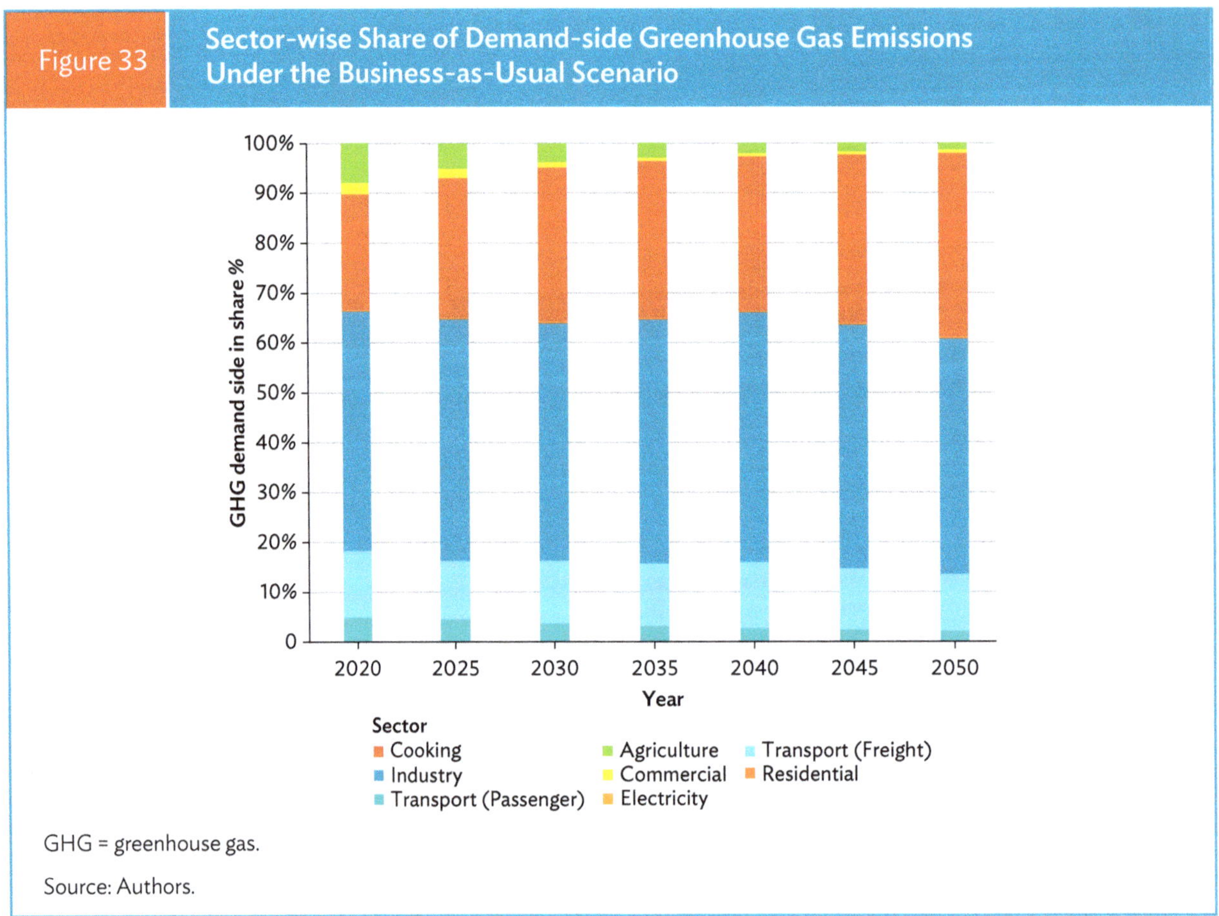

Figure 33 Sector-wise Share of Demand-side Greenhouse Gas Emissions Under the Business-as-Usual Scenario

GHG = greenhouse gas.

Source: Authors.

Low-Carbon Scenario

Under the low-carbon scenario, total GHG emissions reach 124 $MtCO_2e$ in 2030 and 221.7 $MtCO_2e$ in 2050. Compared with the BAU case, this is almost 16% lower in 2030 and 38% lower in 2050.

Supply Side

The total supply-side GHG emissions reach 52.6 $MtCO_2e$ by 2030 and 105.8 $MtCO_2e$ by 2050 (Figure 34). In the low-carbon case, coal contributes only 7% of supply-side GHG emissions in 2030, reducing to 5% in 2050. The share of gas-based emissions in the energy supply is much higher contributing 74% in 2030 and 81% in 2050. Oil-based supply-side emissions account for 20% in 2030 and 14% in 2050 (Figure 34).

Compared with the BAU scenario, total supply-side emissions are 34% lower in 2030 and 51% lower in 2050. In terms of electricity generation alone, emissions are 35% lower in 2030 and 52% lower in 2050.

Total demand-side GHG emissions are projected to reach 71.3 $MtCO_2e$ by 2030 and 115.8 $MtCO_2e$ by 2050 (Figure 35). Industry is the highest emitter with 43% of demand-side emissions in 2030 and 48% in 2050. The share of transport increases slightly from 14% in 2030 to 15% in 2050, while that of residential cooking declines from 40% in 2030 to 36% in 2050 (Figure 36).

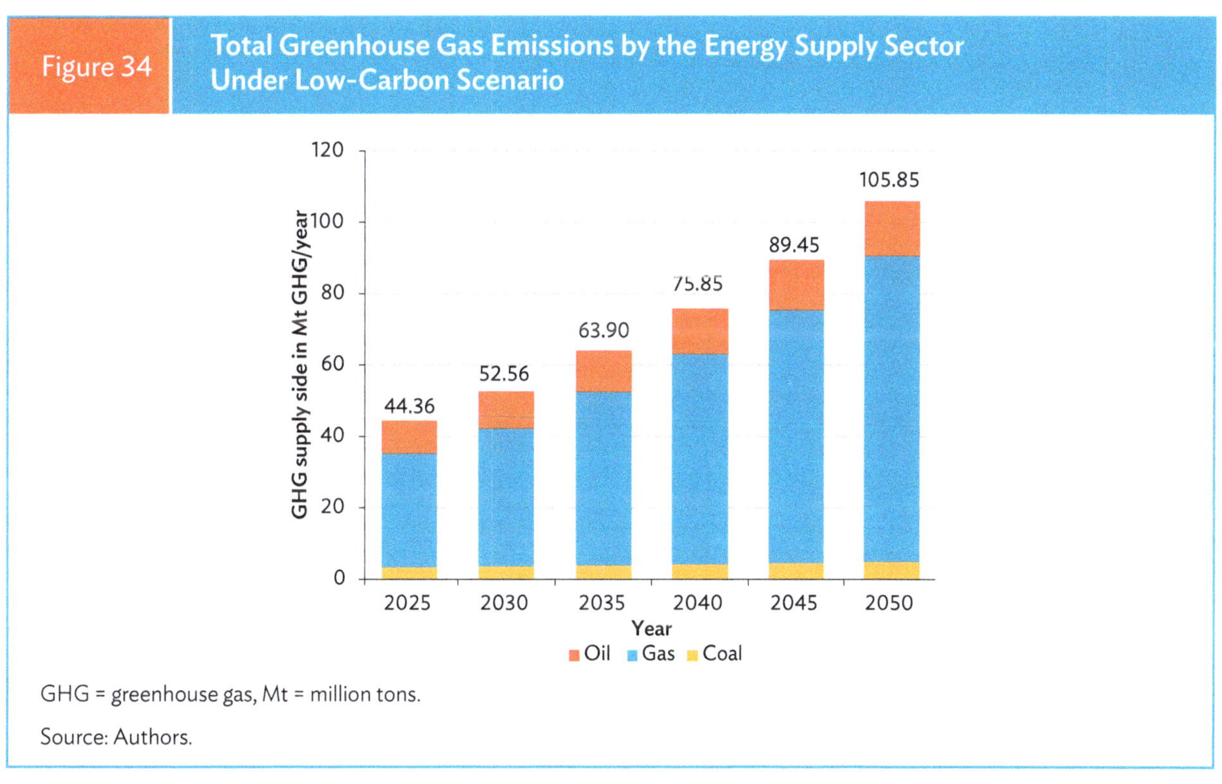

Figure 34 — Total Greenhouse Gas Emissions by the Energy Supply Sector Under Low-Carbon Scenario

GHG = greenhouse gas, Mt = million tons.

Source: Authors.

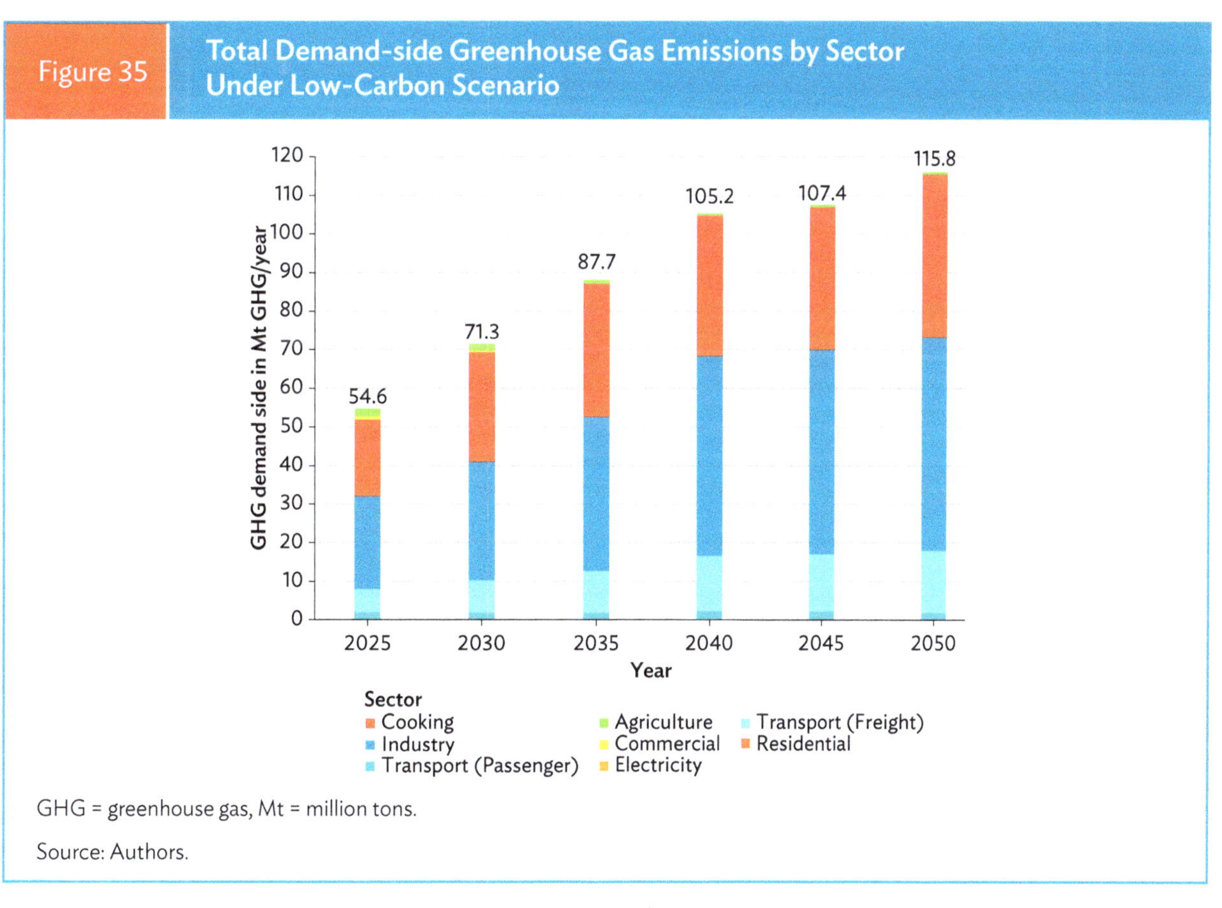

Figure 35 — Total Demand-side Greenhouse Gas Emissions by Sector Under Low-Carbon Scenario

GHG = greenhouse gas, Mt = million tons.

Source: Authors.

| Figure 36 | Sector-wise Share of Demand-side Greenhouse Gas Emissions Under Low-Carbon Scenario |

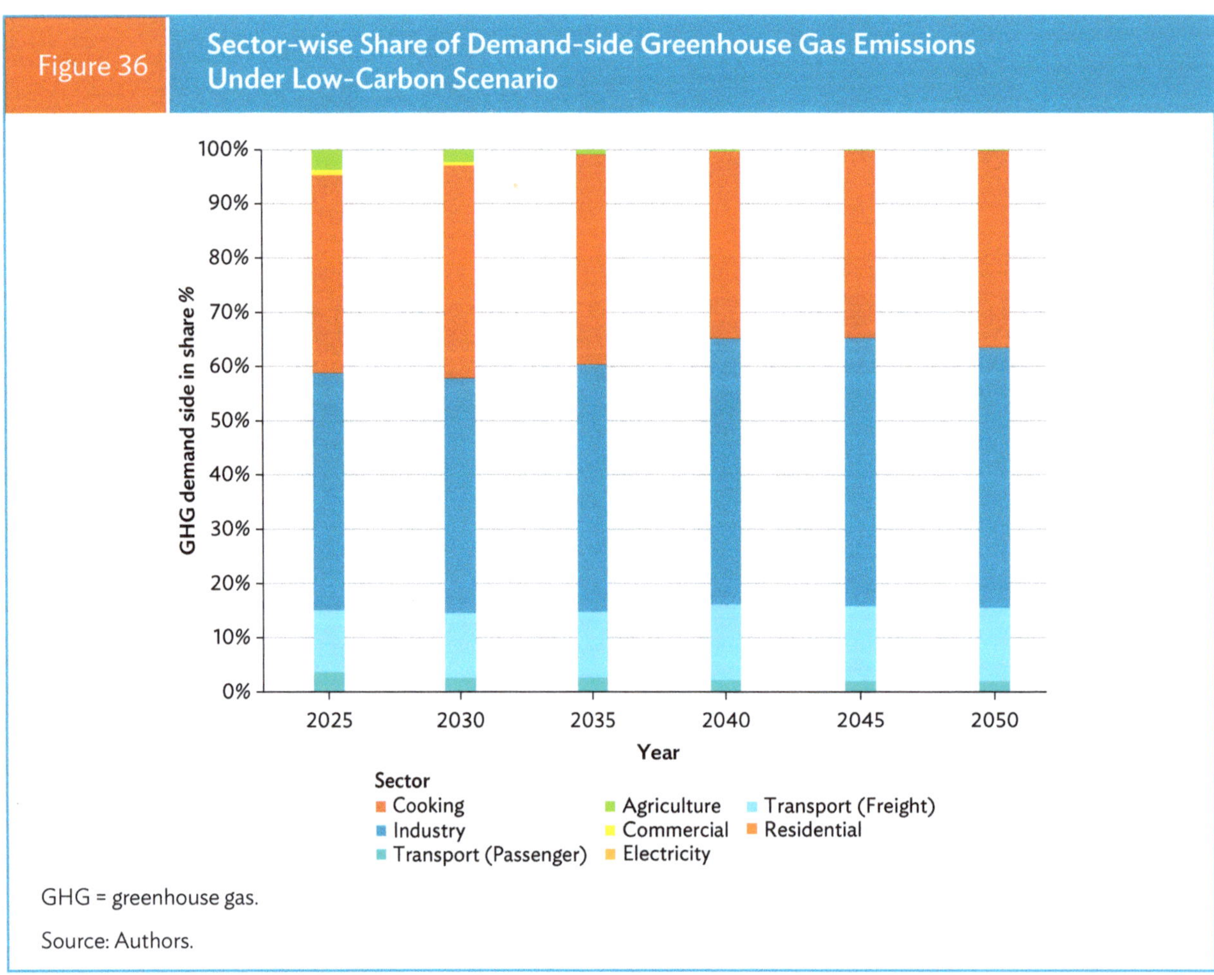

GHG = greenhouse gas.

Source: Authors.

The initial INDC for emissions reduction from 2015 proposed a 5% reduction as the unconditional target for emissions from the power, industry, and transport sectors in 2030 and a further 10% reduction in the conditional target, implying a combined reduction of 20%. The updated commitment proposed a 27.56 $MtCO_2e$ (6.7%) reduction in unconditional and an additional 61.91 $MtCO_2e$ (15.1%) reduction in the conditional scenario, implying a combined reduction of 21.8% (MoEFCC 2015, 2021). The low-carbon scenario developed in this study shows how this can be achieved as by 2030 the cumulative energy-based GHG emissions from the power, industry, and transport sectors would be reduced by 24.1% compared with the BAU scenario. Compared with the BAU scenario, total GHG emissions from all sectors would be about 16% lower in 2030 and 38% lower by 2050, thus helping to achieve a key environmental goal.

3. Alternative Energy Sources and Options for Mitigation and Efficiency Improvements

Scale and Cost of Emissions Mitigation

The model also produces estimates of the scale and cost of emissions mitigation. A typical marginal abatement cost (MAC) curve shows the abatement potential in tons of CO_2 emissions of a project and the cost for achieving those emissions reductions in \$ per ton of CO_2e. Usually, the baseline scenario of the country is considered as the reference, and the emissions savings from various project options are compared with the baseline. For an energy system optimization model, as in this analysis, an alternative approach is often used and the maximum abatement potential of a technology is the emissions abated by replacing the most emission-intensive technology. Dividing the cost of replacing the most emission-intensive technology by the emissions saved gives the unit MAC for the project. In this case the comparison is with the most emission-intensive coal technology (coal subcritical).

For instance, in the case of solar PV plant projects, the model optimizer determines the unit MAC through the following equations:

$$Cost\_abatement_{SolarPV,year} = Levelized\ Cost_{SolarPV,year} - Levelized\ Cost_{Coal(subcritical),year} \qquad \text{Equation 1}$$

The levelized cost is calculated within the MESSAGEix solver taking into consideration the factors such as future capital investment cost, variable cost, and the technical life of a technology and discounting these to the present.

$$Abatement_{SolarPV,year} = (Emission\ Factor_{Coal(subcritical),year} - Emission\ Factor_{SolarPV,year}) * Activity_{SolarPV,year} \qquad \text{Equation 2}$$

$$Marginal\ Abatement\ Cost_{SolarPV,year}\ (\$\ per\ tonne\ CO_2\ abatement) = \frac{Cost\_abatement_{SolarPV,year}}{Abatement_{SolarPV,year}} \qquad \text{Equation 3}$$

where,

Levelized cost covers all investment and operating cost discounted to a base year at a 10% discount rate.

Activity refers to the energy generated or consumed by a particular technology.

From the low-carbon scenario, the key supply-side measure with high mitigation potential is the switch from fossil-based power generation to renewable-based generation, such as solar PV and wind. Key mitigation opportunities on the demand side arise in the transition to gas-based cooking, increasing penetration of efficient appliances in buildings, coal-to-gas transition in industry, and electrification of transport.

Natural gas will have a crucial role in the clean energy transition allowing a phase out of coal utilization in power generation and industry. Deployment of the combined cycle power generation technology not only lowers emissions but also enhances resource utilization efficiency compared to coal subcritical and supercritical power generation. Moreover, due to their high ramping capability and fast response, natural gas power plants can support peak demand and provide load following and/or load balancing services to support the integration of renewable sources. The combined cycle gas turbines can operate at relatively high efficiencies for a minimum load and can ramp up within a few minutes to full load. Emissions per kWh of electricity generation from gas-fired power plants are 60–70% lower than comparable coal power plants. However, while there are many supporting factors to gas-based power development in the country, the high cost of infrastructure required for production and transport is a potential barrier and is considered in the investment cost estimates in the model (see Section 4: Investment Cost and Financing Options).

Based on the low-carbon scenario assessment, total gas-based power generation capacity is estimated to grow from 12.3 GW in 2025 to 23 GW by 2040 and 34 GW by 2050. The combined cycle power plants have a mitigation potential of 716.4 $MtCO_2e$ between 2020 and 2050 with a negative abatement cost of $2.28/$tCO_2e$. A negative MAC implies the levelized cost of the technology is below that of the emission-intensive coal alternative, so the numerator of equation 3 is negative.

Table 11 and Table 12 give the estimated emissions mitigation (relative to the coal option) for key supply and demand side technologies 2020–2050. Gas-based power plants and solar PV offer the largest and most cost-effective reduction in emissions.

Table 11	Mitigation Potential of Key Supply-side Technologies
Fuel (Supply)	**Mitigation Potential in $MtCO_2e$ (2020–2050)**
Combined Cycle Gas PP	716.4
Solar PV	319.8
Imported Electricity	260.3
Nuclear PP	132.5
Combined Cycle Oil PP	109.4
Biomass PP	109.3
Wind	44.3
Hydel PP	14.2
Supercritical PP Coal	5.1
Domestic Gas	2.5
Total	1713.8

$MtCO_2e$ = million tons of carbon dioxide equivalent, PP = power plants, PV = photovoltaic.

Source: Compiled by authors.

Under a low-carbon scenario, some supercritical coal-based power generation capacity will continue. The key enabler for coal-based power generation in the country is its low cost. However, a coal-based power generation system, as well as its emissions issue, offers limited flexibility to the electricity grid in supporting a high integration of renewables in the system due to its low ramping rate. The required flexibility can be provided by other fuels and technology options, like natural gas-based power, pumped hydro, and battery storage technology. Hence the share of coal-based power generation under the low-carbon scenario will be no more than 1.1% by 2050.

Table 12	Mitigation Potential of Key User Sectors
Sector (Demand)	Mitigation Potential in MtCO$_2$e (2020–2050)
Cooking	1045.2
Industry	363.3
Residential and Commercial	323.4
Transport	183.4
Agriculture	55.16
Total	**1970.5**

MtCO$_2$e = million tons of carbon dioxide equivalent.

Source: Compiled by authors.

Investment in solar PV-based power generation is one of key transition opportunities. In the low-carbon scenario, solar PV power generation capacity is estimated to grow from 13.4 GW in 2025 to 48 GW by 2050. Its total emissions reduction between 2020 and 2050 is estimated at 319.8 MtCO$_2$e with an average abatement cost of $1.09/tCO$_2$e. With increasing capacity globally, the cost of solar PV power plant development has come down significantly in the last decade. With the reduction in initial investment costs, its cost of power generation has been reduced to levels, similar to that of coal and hydro-based generation. However, the intermittent nature and variability of solar resources are significant barriers to solar PV capacity addition. While the country has local content requirements for solar power-related manufacturing aimed at protecting domestic manufacturers of solar PV system parts, due to limited demand and manufacturing capacity, the cost of domestic products remains high, which increases overall system cost. Nonetheless mitigation potential is projected as large at a low marginal cost.

Wind-based power generation has significant untapped potential. With a very low installed capacity and immense potential for both onshore and offshore wind farm developments, the country has significant scope for wind-based power generation. In the low-carbon scenario, wind power grows from 0.5 GW in 2025 to 10.3 GW by 2050. The total emissions reduction created is estimated as 44.3 MtCO$_2$e between 2020 and 2050 with an abatement cost of $4/tCO$_2$e (Table 11 and Table 12). As per industry projections, the cost of offshore wind power is expected to come down in the coming years allowing rapid capacity addition. Just like solar, intermittency and variability are barriers and therefore a flexible and smart grid is needed to accommodate a high variable renewable energy share in the power generation fuel mix.

The following sections discuss in more detail the prospects for low-carbon development offered by alternative fuel sources and technologies.

Natural Gas

As of 2017, Bangladesh had 12 trillion cubic feet of natural gas reserves, with one estimate suggesting that proven reserves were equivalent to 10 times annual consumption. While 27 gas fields have been discovered in the Bengal basin, with rising demand, finding more exploitable natural gas reserves is urgently needed (Shetol et al. 2019).

Natural gas-based power plants can be classified into:

- Combined Cycle Natural Gas Power Plant
- Simple cycle gas turbine power plant
- Gas engines

Combined Cycle Natural Gas Power Plant

In a natural gas combined cycle power plant, the natural gas undergoes combustion with compressed air. The hot air-fuel mixture passes over the blades of a gas turbine connected to a generator, which generates power. The second stage involves recovering waste heat carried by the exhaust gases using a heat recovery steam generator to generate steam. The high-temperature steam is used to drive a steam turbine to generate additional power. Large-scale natural gas power plants utilize the combined cycle gas turbine technology due to its higher efficiency and short construction and start-up times, as well as low capital costs.

Simple Cycle Gas Turbine Power Plant

In a simple cycle or open cycle gas turbine power plant, the air enters the compressor and passes through the combustion chamber, where the natural gas undergoes combustion. The high-temperature products of combustion pass over the turbine blades, which rotate. The turbine is connected to the generator that produces power. The turbine also drives the motion of the compressor. The heat from flue gases is extracted for other operations such as district heating, unlike in the combined cycle power plant. Simple cycle gas turbine plants require a shorter time for construction, as well as shorter start-up times. These turbines, however, are limited to around 15 MW capacity, while combined cycle power plants are more energy efficient.

Gas Engines

Gas engines are internal combustion engines, connected to electric generators for power production. Natural gas is used as fuel that undergoes combustion with ambient air inside the spark-ignition engine. The combustion products drive the crankshaft and subsequently the generator. These engines can achieve electrical efficiency of up to 48%. The gas engines have a power generation capacity of up to 150 kW and are used at smaller scales of operation.

These technologies are widely deployed globally. Under the low-carbon scenario, the gas-based power generation is estimated to grow from 12.3 GW in 2025 to 22.8 GW by 2040 and 34.2 GW by 2050, with a mitigation potential of 716.4 MtCO$_2$e between 2020 and 2050 and a negative abatement cost of \$2.28/tCO$_2$e. As natural gas is one of the important low-carbon fuels for the energy transition, it is important to explore and develop new gas reserves as the current reserves are expected to last for only 10–12 years. It is also important to invest in gas infrastructure to handle the growing demand and develop a long-term plan for gas extraction and use. Natural gas-based technologies are technically mature and commercially available and, in the power sector, have the advantage of flexibility, as they are capable of ramping generation up and down quickly as needed, allowing the introduction of a high share of variable renewable energy in the overall electricity mix.

Table 13 summarizes natural gas barriers and enablers for efficiency improvements if the low-carbon scenario is to be implemented.

Table 13	Natural Gas Barriers and Enablers for Efficiency Improvements	
Parameters	**Enablers**	**Barriers**
Cost	• Low cost of power generation. • Government indirectly subsidizes natural gas.	• High cost of infrastructure development especially LNG exploration and pipelines. • Government has steadily increased the natural gas tariff for captive power plants.
Policy and Regulatory Environment	• Slumps in demand due to the coronavirus disease (COVID-19) have made the government shift from prioritizing coal-based to natural gas.	• Reservoir characterization and estimation need to be conducted using advanced methods to convert probable reserves of natural gas to proven reserves of natural gas. • Supply and demand gap has led the government to take several immediate measures to increase the short-term supply of electricity.
Market	• Electricity generation is mostly natural gas based. • Petrobangla account for 99.4% of the supply of natural gas. • Companies from the United States (US) supply over 55% of Bangladesh's domestic natural gas production. • US-origin power turbines currently provide 80% of Bangladesh's installed gas-fired power generation capacity.	• Possibility of natural gas being exhausted within a decade based on supply and demand projections.
Technical and Commercial Performance	• Offshore blocks of the Bay of Bengal may have reserves of untapped gas. • Incentives and tax exemptions to foreign investors are available. • International bidding on 21 offshore gas blocks postponed in 2020 due to outbreak of the COVID-19.	• Depleting resources of old reserves. • Technical know-how about exploration and supply of LNG is limited.
Stakeholder Acceptance	• Increasing interest by the government to find new natural gas reserves.	• Lack of a cost-recovery mechanism increases project uncertainty.
Skill Development	• Abundant cheap workforce is available.	• Lack of technology in the exploration aspects.

Source: Compiled by authors.

Solar

In the low-carbon scenario, solar energy plays an increasing role in power generation, with electricity generation from solar PV increasing from 21.1 TWh in 2025 to 46.4 TWh in 2040 to 75.6 TWh in 2050.

The power generation from solar energy can be divided into the following categories based on the type of installation and connection to the grid:

• Ground-mounted large-scale solar PV
• Rooftop solar PV
• Floating solar power
• Solar PV with battery storage
• Concentrated PV

International best practices currently focus on floating solar and hybrid wind-solar systems.

Ground-mounted large-scale solar PV

The ground-mounted solar PV plant is the most widely used solar technology. Large-scale solar power plants are often ground-mounted due to the lower cost of installation as compared to other options. The power plants consist of an array of solar modules connected to inverters and a centralized monitoring system. The PV modules used in commercial power plants are mono-crystalline silicon (mono-Si), multi-crystalline silicon (multi-Si), and thin-film solar cells. Conventionally, the lower cost and higher efficiency have led to mono-Si substituting the multi-Si solar cells. Among the thin-film solar cells, the most common technologies are cadmium telluride (CdTe), CIGS cells, and amorphous silicon (a-Si) solar cells.

Rooftop solar PV

A rooftop solar power plant is installed similarly to a ground-mounted solar plant, but smaller, and located on the rooftop of commercial and residential buildings. The power generated can be used for in-house consumption and the excess power can be sold to the grid. The system can also be operated in the hybrid mode by combining it with battery storage and at the same time connecting it with the electricity grid.

Floating solar power plant

Floating solar power is emerging as an alternative to ground-mounted solar power plants to overcome the challenge of limited land availability. The floating PV plants are developed on open water bodies such as lakes and canals. The modules are installed with buoyant structures that keep them afloat. The floating PV modules have a better performance due to the natural cooling of the equipment from the water. These plants also reduce evaporation of water, which becomes available for irrigation and drinking purposes. The shading provided by the modules also prevents the development of algae, which are undesirable in drinking water.

Solar PV with Battery Storage

The solar PV with storage is aimed at providing electricity when there is no sunshine and in remote areas. This type of system also ensures reliable power applications where grid power is unreliable or unavailable. A system of this kind can be connected to the grid and is able to operate in stand-alone mode.

Concentrated PV

Concentrated PV technologies use optical concentrators to focus solar radiation. The cells used in these applications have a small dimension with high efficiency. These systems require single and dual axis tracking systems for optimal performance by concentrating the maximum possible amount of light. As the tracking system has active mechanical components, it requires higher maintenance.

The commercial usage of solar power started in the 1970s and innovation in the technologies for harnessing solar energy since then has led to large-scale global adoption. Between 2009 and 2019, the installed capacity of solar power plants globally rose from about 25 GW to 580 GW (Our World in Data). Like the global trend toward solar power, the country is also experiencing the rapid installation of solar PV-based power generation plants. The rapid growth is driven chiefly by the supportive policies for solar power producers combined with growing energy demand and the possibility of rapid installation. The grid-connected solar PV has high emissions reduction potential and is found to be the lowest cost energy supply option with a negative MAC of $3.81/tCO$_2$e. Ground-mounted large-scale solar has had the most rapid growth. At present, the cost of domestic manufacture of solar panels is high with few domestic suppliers. With the increasing demand for solar panels, this provides an opportunity to promote domestic production and provides job opportunities in the clean energy area.

Of the technology categories listed above, current experiences in the country are as follows:

- Ground-mounted large-scale solar PV (rapid growth)
- Rooftop solar PV (ambitious targets, but slow growth)
- Solar PV with battery storage (limited but growing, and supporting energy access)
- Floating solar power (little application but potential success now demonstrated with TRL of 8)

The barriers and enablers for future development of solar photovoltaic if the low-carbon scenario is to be implemented are summarized in Table 14.

Table 14	Barriers and Enablers for Future Development of Solar Photovoltaic	
Parameters	**Enablers**	**Barriers**
Cost	• Decreasing cost of solar farm installation.	• Not as competitive in terms of pricing and market incentives.
Policy and Regulatory Environment	• National Solar Energy Action Plan has three possible scenarios for total solar energy capacity by 2041, BAU (8 GW), medium (25 GW), and ambitious (40 GW). • Government announced plans to install rooftop systems on all educational facilities. • Solar Home System Program by IDCOL. • Government target of generating 10% of the country's total electricity through using renewable energy by 2021.	• Coordination among ministries is missing; procedural difficulties. • Limited budget allocation to renewable energy-based projects.
Market	• Rapid economic growth means rising demand for electricity. • Potential for solar energy is 2,690 MW.	• Lack of incentives for private investors.
Technical and Commercial Performance	• Geographical position gives great potential, absorbs 4.0-6.5 Watt-hour/square meter of solar radiation daily (Halder et al. 2015). • High potential for floating solar projects such as the Kaptai dam, riverbanks, and islands like the Meghna estuary. • Financial support available from the World Bank to fund renewable projects.	• Coastal land and riverbanks are subject to constant erosion and flooding. • Slow technological adoption. • Land shortage for harvesting solar energy.
Stakeholder Acceptance	• Not-for-profit groups such as Grameen Shakti are working toward supplying renewable energy technologies in rural areas.	• Lack of awareness and application of green building policies. • Lack of a better grid infrastructure with enhanced demand and capacity.
Skill Development Requirement	• Low cost of labor and abundant labor supply. • Training planned by Bangladeshi Clean Energy Regulator and the SREDA. • Schneider will support Underprivileged Children's Educational Programme Bangladesh to set up 19 labs to provide skill-based training.	• Shortages of skilled workers in both frontline and mid-level positions.
End of Life	• Solar panels have a long life.	• Currently, no regulations exist for recycling exist.

Source: Compiled by authors.

Wind Power

The country has significant wind energy potential, both onshore and offshore, with over 20,000 square kilometers of land where the wind speeds vary between 5.75 and 7.75 meters (m)/second. This equals a potential of over 30,000 MW. Under the low-carbon scenario, an emissions mitigation potential of 44.3 $MtCO_2e$ between 2020 and 2050 is estimated.

Wind power plants can be classified into the following categories:

- Onshore wind
- Offshore wind: Seabed fixed
- Floating offshore wind

Onshore Wind

Onshore wind turbines are installed on land areas with high wind speeds. The wind turbine generators most widely used are horizontal axis wind turbines. The turbine blades extract the energy from the wind to rotate and drive the shaft, which is connected to an electric generator.

Offshore Wind: Seabed Fixed

Offshore wind power is generally attributed to seabed fixed offshore wind turbines. The wind turbine generators are mounted on a variety of foundations such as monopiles, multi-piles, tripods, and lattice towers. The offshore wind turbines have higher wind resource availability due to little obstruction. However, their installation is complicated and requires sufficient streamlining of the supply chain.

Floating Offshore Wind

Floating offshore wind turbines overcome the structural limitations of offshore wind, which require fixed foundations in ocean depths exceeding 60 m. The floating offshore turbines are supported on a barge, or semisubmersible platforms.

Of these, onshore wind turbine technology is mature and available commercially in the market for rapid deployment (TRL 10). In the case of offshore wind turbines, the technology set up of wind turbine platforms and power evacuation infrastructure is still evolving to support offshore wind farm development (TRL 9). Floating systems are little further behind (TRL 8). Under the various types of installation, technological development and increasing turbine size have led to price reductions globally. In the last decade, the global levelized cost of electricity from a newly commissioned onshore wind energy system fell from $0.089 per kWh in 2010 to $0.039 per kWh in 2020, a 56% reduction (IRENA 2020). Similarly, in the case of offshore systems during the last decade, the weighted average global levelized cost of offshore wind energy has fallen by 48% from $0.162/kWh to $0.084/kWh.

Several wind resource assessment programs and projects are ongoing in the country. The low-carbon scenario suggests that the installed capacity of wind offshore would increase from 0.5 GW by 2025 to 3.9 GW by 2050, while that of wind onshore would increase from 2 GW in 2030 to 6.4 GW in 2050. In terms of MAC, onshore wind (a negative MAC of 1.75/tCO_2e), with a mitigation potential of 25.2 $MtCO_2$ between 2020 and 2050 is considerably lower cost than offshore wind (positive MAC of 11.6/tCO_2e), with a mitigation potential of 19.1 $MtCO_2e$.

While there is significant potential for installation, the country has no domestic manufacturing capacity for wind turbines. Again, as with solar panels, low-carbon development provides an opportunity for domestic production, for example, by initially inviting an international wind turbine manufacturer to set up domestic manufacturing capacity.

Table 15 summarizes barriers and enablers for future development of wind power if the low-carbon scenario is to be implemented.

Table 15	Barriers and Enablers for Future Development of Wind Power	
Parameters	**Enablers**	**Barriers**
Cost	• Wind turbine prices have decreased significantly since 2010. • LNG prices are going up worldwide.	• Subsidization of fossil fuel use.
Policy and Regulatory Environment	• Government target of generating 10% of the country's total electricity through using renewable energy by 2021. • Various tax exemptions available to promote investments in renewable power.	• Coordination among ministries is missing and there are procedural difficulties. • Limited budget allocation to renewable energy-based projects.
Market	• First major wind power plant project with a capacity of 55 MW approved by Cabinet in 2020. • Government initiative to set up three wind power plants in recent years. • Several wind power independent power producers (IPPs) are under development. • Increasing share of electricity in energy consumption .	• Lack of investment due to low competitiveness with global plants and fossil-based sources in Bangladesh.
Technical and Commercial Performance	• Over 20,000 square kilometers of land where the wind speeds vary between 5.75 and 7.75 meters per second. This equals a potential of over 30,000 MW. • Several wind resource assessment programs are ongoing.	• Implementation delay in earlier projects due to failure of the bidder. • Limited land availability.
Stakeholder Acceptance	• Envision Energy has shown interest in setting up plants at Mongla and Inani in 2019. • Increasing public awareness due to air pollution in cities.	• Lack of information about the credibility of developers.
Skill Development Requirement	• Low cost of labor and abundant labor supply. • Training planned by clean energy regulator, SREDA.	• Unavailability of sufficiently trained engineers and technicians for manufacturing, design, construction, operation, and maintenance of wind power plants.
End of Life	• Wind turbines have long lives. During their operational lifetime, the improvement in technology is significant. As a result, wind farms can be repowered with new wind turbines at the same location.	• There are standards for recycling of wind turbines. • Wind turbines made of composite materials are difficult to recycle.

Source: Compiled by authors.

Coal

Historically, coal has been one of the cheapest sources of energy, which can be easily stored and transported. However, globally awareness of climate damage has led to pressure to phase out the use of coal. The country has never had a significant dependence on coal as an energy source, with less than 2% of electricity supplied from coal-based plants in 2018. In June 2021, the government took the decision to scrap plans to build 10 new coal-fired power plants, which would have accounted for 8,451 MW of power. The low-carbon scenario allows for some small use of coal (accounting for 1% of power generation in 2050), but only from the more efficient and less polluting supercritical power plants.

Coal plants can be classified into the following categories:

- Sub-critical coal power plant
- Supercritical coal power plant
- Ultra-supercritical coal power plant
- IGCC power plant

Sub-critical, Supercritical, and Ultra-supercritical Coal Power Plants

Power generation requires pulverized coal, which is combusted with air. The heat generated is used to produce steam that drives the steam turbine and generator. The pressure and temperature conditions of the water decide the class of coal power plants. The water above a critical point (374 degree Celsius [°C], 221 bar) is indistinguishable from steam. The boilers in sub-critical power plants have a temperature and pressure well below the critical point. In supercritical boilers, at pressure and temperature above the critical point, the water can remain in liquid form even at temperatures well beyond its boiling point at atmospheric pressure. The higher temperature enables higher efficiency of conversion to mechanical power in the steam turbine and higher electrical efficiency. Supercritical boilers can reach a temperature up to 565 °C. The ultra-supercritical boiler technology has even higher temperature and pressure reaching 580–600 °C and 241–310 bar, which enables higher conversion efficiencies.

Integrated Gasification Combined Cycle (IGCC) Power Plant

In an IGCC power plant, the coal is first converted to syngas (CO+hydrogen [H_2]) in a high-pressure gasifier. The syngas is then used in a combined cycle. The syngas is purified by removing impurities, such as sulfur dioxide, mercury, CO_2, and particulate matter. The purified syngas then passes through the gas turbine where it undergoes complete combustion for power generation. The waste heat from the flue gases as well as the gasification products is extracted through the heat recovery steam generator for additional power generation. An IGCC offers lower emissions compared to pulverized coal-based technology, and a high CO_2 concentration of emissions, resulting in more in efficient CCUS applications.

Figure 37 gives the comparison of the thermal to electricity conversion efficiency of different coal technologies. While these are the standard efficiency levels, an actual plant level efficiency may vary due to other ancillary processes associated with the main power generating units. By these estimates the most efficient coal technology ultra-supercritical is about 10 percentage points more efficient than the least efficient, coal sub-critical. All these technologies are widely deployed globally.

The low-carbon scenario has the coal power sector growing modestly from 0.7 GW in 2020 to 0.85 GW by 2050 (Figure 38). The reason for continued requirement of coal for electricity generation is because of a lack of alternative technologies to balance electricity demand. It is unlikely that the installed capacity of renewables

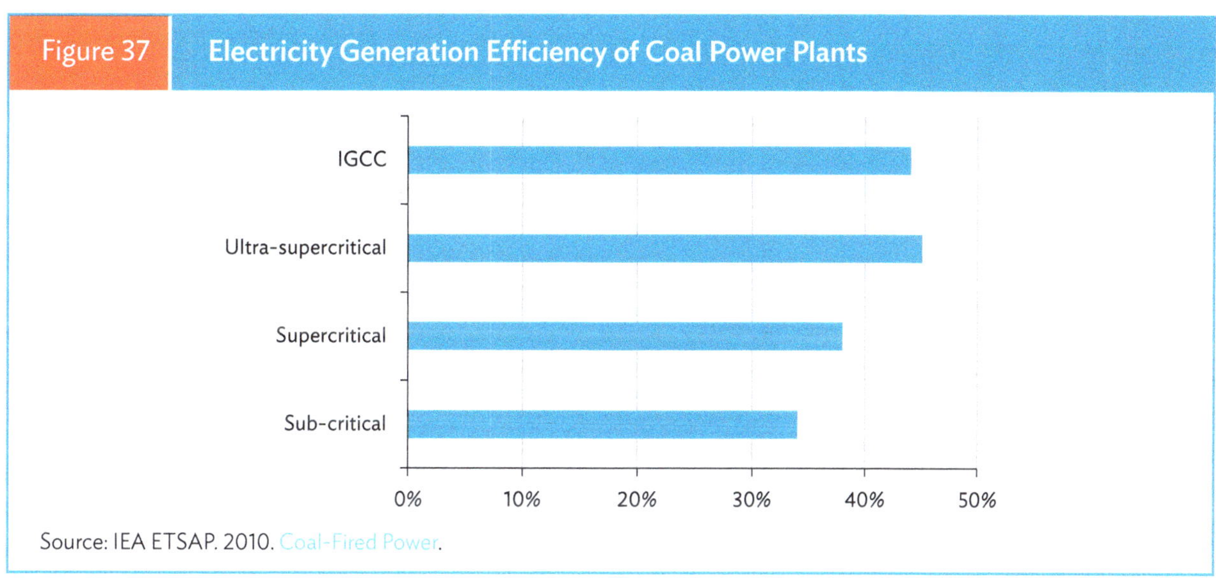

Figure 37   Electricity Generation Efficiency of Coal Power Plants

Source: IEA ETSAP. 2010. Coal-Fired Power.

would exceed the present scenario estimation of around 64 GW in 2050, due to the limitations on natural resource potential. Thus, even though the cost of renewables might be declining, some fossil-based power generation is still required in the low-carbon scenario as a limited, additional source. The low-carbon scenario estimates that power generation using coal supercritical power plants have a mitigation potential of 5.1 $MtCO_2e$ between 2020 and 2050, but with a high abatement cost of $17.85/$tCO_2$e.

Table 16 summarizes barriers and enablers for future development of efficient coal-based power if the low-carbon scenario is to be implemented.

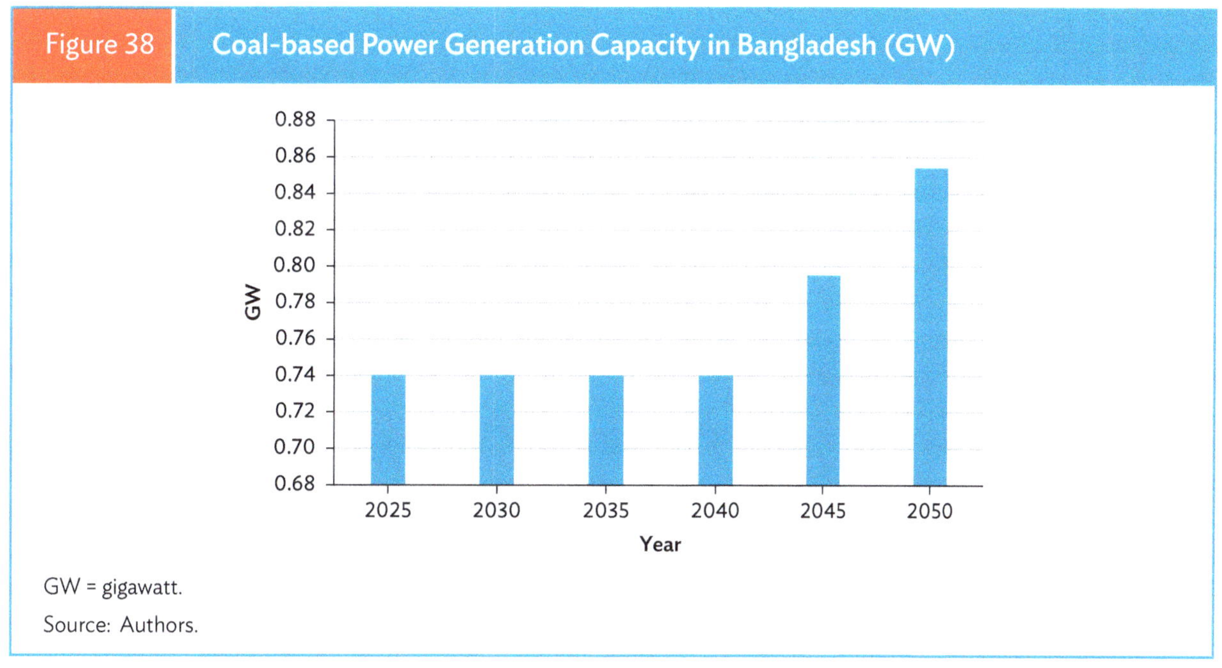

Figure 38   Coal-based Power Generation Capacity in Bangladesh (GW)

GW = gigawatt.
Source: Authors.

Table 16	Barriers and Enablers for Future Development of Efficient Coal-based Power	
Parameters	**Enablers**	**Barriers**
Cost	• Low cost of generation	• Cost of competing energy sources such as wind, solar, and natural gas is reducing. • Increasing coal prices.
Policy and Regulatory Environment		• Plans to build 10 coal-fired power plants dropped. • Target to generate 40% of power from renewable energy by 2041.
Market	• Three large coal power plants, amounting to 5 GW of capacity are connected to the grid.	• Government wishes to discourage the use of coal on environmental grounds.
Technical and Commercial Performance	• Largest coal power pipelines in the world, a total of 29 power plants amounting to 33.2 GW of capacity, but 90% of them are under review to find alternatives. • Investment that offers stable returns.	• Equipment needed for coal-fired plant and coal needs to be imported. • Concern after government's review of coal-fired plants.
Stakeholder Acceptance	• Ultra-supercritical technology is being adopted to make coal power plants more environmentally friendly.	• Concerns about coal-based power plants expressed by climate activists may influence policy.

Source: Compiled by authors.

Road Transport

The transport sector largely consumes oil-based energy and has significant scope for emissions mitigation, with electrification offering the most promising route to decarbonization. The electrification of road transport will be through electric vehicles either as battery electric vehicles or as plug-in hybrid electric vehicles (PHEV). The electric vehicle has an electric powertrain with electric motors, which draw power from a battery pack. For optimal performance, the electric vehicle is also equipped with a battery management system, motor controller, power electronic converters, and transmission system. PHEVs, additionally, also have an internal combustion engine powertrain system, which can operate to boost performance or when the battery charge is depleted. The most common modes of electrified transport are:

- Electric two-wheelers
- Electric cars and taxis
- Electric buses

The successful deployment of electric vehicles requires the installation of an extensive charging infrastructure, either domestic or public charging. Globally, there are number of charging standards adopted such as combined charging system Type 1 and Type 2 in EU and North America, GB/T in the People's Republic of China, CHAdeMO in Japan, and Tesla's supercharging system. Each of these offers different functionality, charging loads, and communication protocols.

Even without electrification, emissions from the freight sector can be improved by upgrading emission norms and deploying proper testing measures. Countries mostly pattern their emission policies on European regulations and the associated mandates for clean, low-sulfur fuels. By adopting the Euro 6/VI vehicle emissions standards, it is estimated that the country can achieve up to a 99% reduction in the emissions of pollutants like fine particulate matter (PM2.5), thus reducing the risk of diseases such as ischemic heart disease, lung cancer, strokes, and asthma.

Electric vehicle technology is commercially mature, the ownership cost of vehicles is falling, and the driving range is improving. However, currently the penetration of electric vehicles in the country is limited. The government aims to increase the share of electric vehicles to at least 15% of all registered vehicles by 2030. The Draft Automobile Industry Development Policy 2020 proposes providing several incentives and tax benefits for electric vehicle manufacturing and assembly. But there are significant barriers in transitioning to electric vehicles, particularly limited infrastructure availability, and currently the market is dominated by oil-fueled vehicles. There has been some interest from the private sector in developing the electric vehicle market, but the country still lags in terms of policy reforms and market development for electric vehicles. On the emissions norms front, the proper facilities to enforce the current vehicular emissions standards are not available. As per the 2017 Third Pole report, 75 out of 76 offices responsible for issuing fitting certificates lacked the equipment to examine vehicular emissions.

Nonetheless, the low-carbon scenario is based on assumptions about increased electric vehicle penetration in both passenger and freight transport. It estimates that with efficiency gains due to electrification and low-carbon electricity supply, emissions mitigation from electric vehicles will be 183.4 $MtCO_2e$ between 2020 and 2050, with the highest emissions mitigation potential in the electric three-wheeler segment at 107.7 $MtCO_2e$. The MAC of the electrification of transport in three-wheelers is estimated to be $ 2.79/tCO_2e by 2050.

Table 17 summarizes barriers and enablers for future development of electric vehicles if the low-carbon scenario is to be implemented.

Table 17	Barriers and Enablers for Electric Vehicles	
Parameters	**Enablers**	**Barriers**
Cost	• Price of e-motorbike is now close to that of standard combustion motorbike. • Cost of vehicles can be brought down by local manufacturing and assembly.	• High upfront costs for e-cars.

Policy and Regulatory Environment	• Increase the share of electric vehicles to at least 15% of all registered vehicles by 2030.
	• The Draft Automobile Industry Development Policy 2020 proposes:
	  – Tax exemptions and holidays for investment in assembly facilities for energy-efficient vehicles.
	  – Tax incentive program for assemblers and components makers of electric vehicles.
	  – Permit the import of electric vehicle with prior certification for the recycling and disposal of lithium-ion batteries.
		• Limited supporting infrastructure (14 electric vehicle charging points with a total capacity of 278 kW).
		• Limited national policies.

Market	• High demand expected in the near future. • Positive global trend. • Three-wheeler and two-wheeler market can be dominated by electric vehicles with motorized rickshaws (240,000) and e-bikes (1 million).	• Infant market. • Commercial direct current (DC) fast charging has not yet been introduced.

continued on next page

Table 17	*continued*	
Parameters	**Enablers**	**Barriers**
Technical and Commercial Performance	• Only 40% of power generation capacity was utilized in 2019-20, EVs can solve the overcapacity problem (Annual Report 2019-20—Bangladesh Power Development Board). • Investment growing in the electric vehicle sector, BDAuto has invested \$200 million to locally manufacture electric vehicles. • Walton, Runner, Akij, and Duranta are planning to distribute locally assembled two-wheeler electric vehicles.	• Low availability of e-motorbikes. • High-load commercial usage is not feasible at the current electric vehicle charging points as they are operating using solar energy.
Stakeholder Acceptance	• Investors such as Indian automaker Omega Seiki and local auto manufacturer BDAuto are planning to manufacture three-wheeler electric vehicles.	• Lack of awareness regarding how electric vehicles work. • Local market of electric vehicles is not sizable enough to attract investment. • Hybrid vehicles have better acceptance than fully electric.
Skill Development Requirement	• Low cost of labor.	• Challenge to attract manufacturers to develop electric vehicles locally.

Source: Compiled by authors.

Agriculture

Electricity is often used for groundwater extraction and its use will increase as agricultural pumps become affordable for small farmers. At present, 55% of the cultivable area is under irrigation, mainly by pumped groundwater fueled by diesel oil, although there is expected to be a significant increase in electric and solar pumps in future. Due to lack of knowledge among farmers, the pumps used for agriculture are often oversized. Consequently, they run at suboptimal loading, which leads to higher energy consumption. The key low-carbon alternatives are:

• Electric pumps
• Solar pumps

Electric Pumps

Electric pumps are equipped with electric motors, usually single- or three-phase induction motors, which drive the centrifugal pumps. Electric pumps have higher operating efficiency as compared to diesel pumps and new pumps have a controller, which allows control of the rate of pump speed and thus, of the water flow. Also, they are equipped with water level sensors, which turn off the pump when the level reaches below a set limit, thus, preventing the dry run of pumps.

Solar Pumps

Solar pumps are electric pumps connected to a solar PV system. The PV modules convert the solar energy into electricity, which is used to drive the electric pumps. The operating cost of these pumps is low as the primary energy is practically available for free. These systems are available in two types:

• Off-grid solar pumps: Off-grid systems are stand-alone solar-powered pumps, which operate during the availability of solar energy. These systems effectively satisfy the demand during sunny days when the requirement is highest. Usually, 6–7 hours of daily sunlight is sufficient to provide the water needed. The main advantage of these pumps is that they do not depend on the availability of grid electricity. Also, the issues of power quality are eliminated. Furthermore, the off-grid solar pumps help in water conservation.

- Grid-connected solar pumps: The on-grid solar systems can be operated when the sun is shining, but also during the night when needed. They offer the advantage of selling any excess electricity generated to the distribution company as additional revenue for farmers.

Apart from these options, improvement of efficiency of existing pumps, as well using correct sizes of pump can reduce the overall emissions impact of agricultural irrigation. The electric irrigation and off-grid irrigation technology are commercially mature and with rapid technological advancement the ownership cost of an irrigation system is falling.

Currently the country primarily relies on diesel-based irrigation pumps with their share of the market being more than 50% in 2020. Electrification of these pumps provides an opportunity to drive efficient energy utilization in the sector, as well as to reduce emissions. In addition, the off-grid solar PV-based irrigation pump sets can be installed to replace the old electric pumps, contributing further to mitigating emissions. These shifts are built into the low-carbon scenario. Under the low-carbon scenario, deployment of solar irrigation pumps outpaces that of electric pumps, with diesel pumps almost totally phased out. By 2050, the country is expected to have a total of more than 13.4 million water pumps, of which 88% will be solar pumps and 11% electric. The model estimates emissions reduction due to greater use of electric pumps between 2020 and 2050 at 29.28 $MtCO_2e$, while that for off-grid solar pumps is estimated at 25.45 $MtCO_2e$. The corresponding marginal costs of abatement indicate a negative marginal cost of $7.56/$tCO_2e$ for electric pumps and a positive abatement cost of $4.68/$tCO_2e$ for off-grid solar pumps.

Solar irrigation systems primarily cover three crop seasons, leaving more than 50% of solar power unused. This surplus energy can be injected into the grid, contributing to the overall energy supply. Farmers can use the on-grid power through feed-in tariffs or net metering schemes. In such situations, however, when replacing diesel pumps, new transmission lines will be needed for net metering systems. For a sustainable and low-carbon growth in agriculture, electricity and solar-based irrigation pump sets need to be promoted to speed up their uptake.

Table 18 summarizes barriers and enablers for future development of efficient irrigation pumps if the low-carbon scenario is to be implemented.

Table 18	Barriers and Enablers for Efficient Irrigation Pumps	
Parameters	**Enablers**	**Barriers**
Cost	• Operating and maintenance cost for solar pumps is low.	• High initial costs of materials and installation cost of solar pumps.
Policy and Regulatory Environment	• For solar-powered irrigation pumps, IDCOL provides grants of 50% of the total project cost with 35 percent as soft loan at an interest rate of 6% and 10 years as tenure. • 8th Five-Year Plan focuses on good agricultural practices and environmental sustainability. • Bangladesh Delta Plan 2100 offers major opportunity for climate-compatible development in agriculture.	• Possible political resistance to the necessary policy change. • Shrinking average farm size and diseconomies of scale for marginal farmers need to be addressed.
Market	• Agricultural use of electricity is growing at a faster rate in comparison with domestic and industrial uses. • Huge potential of solar irrigation system.	• Installation, operation, and maintenance of solar pumps is difficult.

continued on next page

Table 18	*continued*	
Parameters	**Enablers**	**Barriers**
Technical and Commercial Performance	• Irrigation is limited, of poor quality and with poor maintenance. • Nonrenewable chemical fertilizers and pesticides make up the largest share (68% to 84%) of total energy inputs in southwest coastal Bangladesh (Talukder and Hipel 2019). • Higher energy efficiency of electric pumps. • Reduced emissions with solar pumps. • Solar pumps prevent over-exploitation of water resources. • Modern pumps allow effective control and demand-side management of power. • Several international organizations, as well as IDCOL, support financing initiatives for renewables.	• Lack of control over irrigation timing for individual farmers because of joint farming.
Stakeholder Acceptance	• Farmers with diesel pumps can shift to electrified and solarized pumps due to reduced cost of operation. • Bengal Renewable Energy Limited with the financial support of IDCOL has installed 63 large-scale solar irrigation pumps.	• Coordination problem among farmers where there is joint farming because of fragmented landholdings and small plots. • Competition for land use between aquaculture and agriculture.
Skill Development Requirement	• The International Finance Corporation (IFC) is actively promoting climate-resilient agricultural practices in Bangladesh.	• Lack of knowledge among farmers regarding best practices in agriculture.
End of Life	• Solar and electric pump systems have a longer life than alternatives.	• Disposal of solar modules at the end of life will require establishment of appropriate solar waste collection mechanisms.

Source: Compiled by authors.

Buildings

Globally, almost 40% of the total CO_2 emissions come from buildings and construction and so with rising incomes this is an important area for improvement. Overall, the building sector can be divided into residential and commercial buildings. Energy demand from the residential sector arises primarily from electricity use in domestic systems, such as water heaters, air-conditioning and cooling, household appliances, and cooking. Energy demand from the commercial sector arises mainly from electricity use for operating systems such as water heaters, street lighting, water pumping, and cooling. Being in the tropical zone the country does not require heating in offices and therefore within the category HVAC, air-conditioning is the major energy consumer.

Creating energy-efficient and sustainable buildings is currently a government objective. The Building Energy and Environment Rating System for Bangladesh was drafted in 2018 by the SREDA. However, the country still lags those of European standards, for example, where the energy efficiency of buildings is concerned.

Apart from cooking, which is discussed separately, energy consumption in commercial and residential buildings, is in the following areas:

• Lighting
• Thermal comfort—HVAC system, fans, air coolers, and heaters

- Water heating—electric geysers, rods, and gas heater
- Electrical appliances—refrigerators, kitchen appliances, TV and entertainment systems, computers, washing machines, dish washers, etc.

Lighting

Incandescent and fluorescent lamps are used widely for lighting purposes. However, with technological development and incentive government schemes, internationally increasingly LED lights are being adopted. LED lights consume less power for the same amount of illumination and have a long life. For a successful low-carbon transition, it is essential that they fully replace the alternatives. Energy consumed in lighting applications can also be significantly reduced using intelligent building automation systems, which using sensors turn lights on and off automatically based on occupancy. In future these applications will be important in new commercial buildings.

Thermal Comfort

Technologies used for thermal comfort include HVAC systems, fans, air coolers, and heaters. In the country the Energy Star Level rating for these consumer appliances has had an impact on the buying habits of consumers. Also, the efficiency of HVAC systems has been improved significantly in recent years. Inverter-based DC air-conditioners are particularly useful for residential applications in the country, as they can allow for fluctuating power supply conditions. In commercial buildings, the energy performance of these systems can be improved substantially with the use of building automation systems. The use of fans should also be encouraged along with HVAC systems to enhance cooling and the distribution of air flow. Air coolers and heaters have been popular in Bangladesh, primarily due to their relative affordability. However, awareness about their energy performance has remained low and many consumers still use inefficient devices.

Water Heating

Water heating is one of the important applications for residential, as well commercial buildings such as hotels. Commonly, water heating is carried out using electric water heaters, LPG cylinders, natural gas, or biomass burners. Alternative technologies such as rooftop solar water heating systems are also commercially available. For commercial applications, solar heating systems and electric heat pump technology can significantly reduce energy consumption.

Electric Appliances

Refrigerators, washing machines, TV and entertainment system, computers and mobiles, blenders, and dish washers are the most common electric appliances used by residential consumers. The adoption of these products in the country is growing with rising income levels, while the energy efficiency of these consumer appliances is also increasing with technological development. Ratings under the Energy Star labeling system helps to create awareness among consumers concerning their respective environmental impacts.

To limit energy demand growth, it is essential to increase the penetration of energy-efficient equipment in domestic and commercial buildings and to improve building standards. The government has set the target of 20% energy efficiency improvement for the sector by 2030. However, lack of public awareness around policies and incentives available are major barriers. Different building regulations are in place throughout the country, such as the Building Construction Rule, 2008 that is applicable for Radjhani Unnayan Kartripakkha (RAJUK) and Chattogram Development Authority only. All other development authorities (like those for Khulna, Rajshahi, and Cox's Bazar Development Authority) and the *pourashavas* (municipalities) are under the purview of the Building Construction Rule, 1996. SREDA is the public sector body responsible for formulating policies and a framework toward implementation of energy efficiency in buildings. Its EECMP from 2016 contains the 20% target noted above. There is further scope for the energy saving through additional measures, which must be considered by SREDA in future cycles of the EECMP.

The low-carbon scenario assumes a fairly rapid acceptance of improved standards and between 2020 and 2050 projects emissions reduction of 209.64 $MtCO_2e$ from the residential sector through lower use of electricity with a negative MAC of $199.98/$tCO_2e$. Total emissions reduction from the commercial building sectors is estimated at 113.7 $MtCO_2e$ with a positive MAC of $8.26/$tCO_2e$.

Table 19 summarizes barriers and enablers for efficiency improvement in construction if the low-carbon scenario is to be implemented.

Table 19	Barriers and Enablers for Efficiency Improvement in Construction	
Parameters	**Enablers**	**Barriers**
Cost	• Cost savings due to energy conservation.	• High upfront cost of investment. • Benefits are not directly visible.
Policy and Regulatory Environment	• LEED Certification in Bangladesh (650 registered buildings so far).[a] • Building Energy and Environment Rating (BEEER). • SREDA is mandated to implement energy efficiency programs and to formulate policies, rules, regulations, and guidelines. • Target is 20% energy efficiency by 2030.	• Lack of awareness regarding policies and incentives. • Different building regulations in different parts of the country.
Market	• Energy efficiency labeling by SREDA sets the performance standards for building appliances.	• Global best-practice technologies and materials for energy conservation in buildings are often not used.
Technical and Commercial Performance	• Potential for reduced energy linked emissions since buildings consume a major share of national total energy. • Central Bank has instructed banks and financial institutions to offer support for green initiatives that follow policy guidelines.[b]	• Benchmarking energy use requires expertise and energy auditors; owners often consider this as an additional cost and are skeptical about potential energy and monetary savings.
Stakeholder Acceptance	• High-end consumers are likely to create demand for building automation and smart systems.	• Inadequate knowledge about building energy performance and certification of developers and architects. • Consumers lack knowledge about energy consumption of household appliances.
Skill Development Requirement	• Training and certification are organized by SREDA in support of United Nations' program Climate Technology Centre Network (CTCN)[c] for low-cost climate-resilient housing.	• Trained workers required for implementing energy efficiency measures. • Limited skills to operate the energy simulation program.
End of Life	• Equipment such as LED lights have significantly longer life than alternatives.	• Effective segregation of materials and their sustainable disposal is a challenge.

[a] SREDA. 2020. Bangladesh Experience in Buildings Energy Efficiency Building Energy Efficiency & Environment Rating (BEEER) System for Bangladesh.

[b] A. Bhattacharjee. 2021. Energy Efficiency and Green Building Pathways. *New Age.* 23 February.

[c] UN Environment Programme. 2018. Development of a certification course for energy managers and energy auditors of Bangladesh.

Source: Compiled by authors.

Cooking

Nationally, the main fuel for cooking is predominantly biomass providing over 80% of cooking fuel. However, in urban areas the major source of fuel for cooking is LPG and a small proportion of dry biomass, whereas in rural households most cooking is with biomass. LPG penetration is expected to increase in future, especially in urban areas. The lack of reliability of power supply affects the public acceptance of electric cooking, but in future there is an opportunity to use rooftop solar with electric cooking.

The low-carbon scenario finds that changing household fuel use for cooking has the largest single impact on energy efficiency among the demand sectors, so there is an urgent need to shift away from biomass to electric and gas-based cooking. Of the electric technologies available, the most energy efficient are induction cookers, which heat the surface of the cooking pan only, leaving little energy loss when compared to a heat conduction process in other types of cooking appliances. In conduction heating, for example, energy efficiency can be adversely affected, if there is any weakness in the design or workmanship of the appliance, leading to reduced heat conduction from the heating element to the cooking pan. Usually, the body of the hotplates are made from steel and there is significant conduction from the heating element to the metal body that causes additional energy loss by convection or radiation.

Modern electric stoves are more efficient, emit less emissions and are safer than traditional cook stoves or three-stone-fires and embody widely available technology. As per the CAP for Clean Cook Stoves 2030, the aim is to achieve 100% clean cooking by 2030. Future government renewable energy initiatives are likely to be channeled through SREDA and awareness campaigns and promotional schemes focusing on electric-based cooking and use of LPG are already in place. However, there are still obstacles preventing many people from switching completely to this mode of cooking, particularly the lack of reliability of power supply. To have a large-scale penetration of electric cooking, government initiatives and supportive policies will be required to support the transition to clean and energy-efficient cooking.

The low-carbon analysis adopts a positive view of this and suggests a significant shift that can be achieved through the increasing penetration of the electricity grid in the country, and the availability of gas for cooking through an expansion of the extensive national reserves combined with policy support. It projects an emissions reduction from cooking of 1,045.2 $MtCO_2e$ between 2020 and 2050, with a weighted average MAC of $0.43/$tCO_2$e. Cooking with gas as fuel has the highest emissions reduction potential of all the cooking fuel options at 781.65 $MtCO_2e$ with a MAC of $ 0.23/tCO_2e.

Table 20 summarizes barriers and enablers for energy efficiency in cooking if the low-carbon scenario is to be implemented.

Table 20	Barriers and Enablers for Energy Efficiency in Cooking	
Parameters	**Enablers**	**Barriers**
Cost	• Operating costs of induction stoves are lower than electric filament and LPG stoves.	• Innovative business and financing models needed to reduce distribution costs and increase affordability. • High upfront cost for biogas plant, while biomass is cheap and readily available in rural areas.
Policy and Regulatory Environment	• National Action Plan for Clean Cooking in Bangladesh (2020–2030) aims to achieve 100% clean cooking access by 2030. • Tax exemptions for LPG imports. • IDCOL has provided subsidies to establish biogas plants across the country. • Market Development Initiative for Bondhu Chula.	• Incentives needed for the private sector to invest in alternative technologies and fuels, such as ethanol, pellet-based intercharge system, and biogas. • Land availability for biogas. • Policy uncertainty regarding LNG vs. LPG.
Market	• Almost 80% of households (of a total 35 million households) lacked access to clean cooking alternatives (2018). • 75% of households nationwide use firewood and other biomass fuels, as do 55% of clean stove users in rural areas. • Annual growth of 8–10% in LPG use nationwide.	• Affordability issues. • Lack of reliability of power supply affects acceptance rate of electric cooking.
Technical and Commercial Performance	• Bangladesh accounted for 47% of total tracked clean cooking investments in 2018, mainly from two large projects financed by the World Bank and the Green Climate Fund.	• Lack of access to alternative technologies. • Lack of short-term finance for capital investment, developing distribution chains, and working capital.
Stakeholder Acceptance	• Opportunity to combine rooftop solar with electric cooking to get more than 50% of cooking energy from rooftop PV.	• Lack of awareness of health benefits and other sociocultural reasons.
Skill Development Requirement	• Awareness campaigns and promotional schemes are in place with government support through IDCOL.	• School curriculum and pedagogical methodologies need to be used to increase awareness around clean cooking. • Low awareness of other technologies, such as ethanol and pellet-based improved cook stoves and electric-based cooking.

Source: Compiled by authors.

Industry

Industry is one of the key growth sectors of the economy, but while industrialization contributes to economic growth it also contributes significantly to emissions. The main energy-intensive industries in the country have been identified as infrastructure, textiles, and pharmaceuticals with an overall end-user energy consumption of 6.8 Mtoe (2017) (PWC estimate). Industry in general uses natural gas as its primary source of energy and in 2015 took around 17% of the total natural gas supply (Malek et al. 2015).

In the model the calculation for industry's total energy demand considered the range of fuels consumed in the sector:

- Solid hydrocarbons: solid fuels are mainly coal, lignite.
- Liquid hydrocarbons: liquid fuels are mainly kerosene, diesel, bio ethanol, bio diesel.
- Gaseous hydrocarbons: gaseous fuels are mainly LPG, natural gas.
- Biomass: this category covers raw biomass as well-derived biofuels such as bio ethanol, bio diesel, and biogas.
- Electricity: electricity is categorized as a commodity covering grid-connected electricity that can be from both conventional and renewable-based, power plants.

Traditional Boiler Fuels

At present, producers largely use coal- and oil-fired boilers for steam or hot water production. These are either water-tube boilers or fire-tube boilers. In both, the fuel undergoes combustion and transfers the heat to the water to provide hot water or steam as per the requirement. These are likely to remain dominant and be responsible for most emissions in the industrial sector. However, to mitigate their environmental impact, energy efficiency or technology improvement measures can be adopted. Some of these measures are:

- Combustion control optimization
- Flue gas heat recovery
- Soot blower optimization
- Chemical absorption for post-combustion carbon capture
- Oxy-fuel combustion
- Chemical looping combustion
- Physical absorption of CO_2 from gasification of coal to syngas in water-gas-shift reaction

Alternatively, fuel-switching to low-carbon alternatives such as natural gas, biomass, or H_2 can be applied in these boilers. Gas-fired boilers have an added advantage of flue gas condensation for steam recovery, which can improve the overall efficiency of operation. Biomass is a renewable and carbon-neutral source of energy. One of the major advantages of biomass is that, apart from being used independently in a biomass boiler, it can be a substitute fuel use in a coal-fired boiler with little modification. Continuous usage of biomass for heat generation requires establishing a reliable supply chain for collection and transportation of biomass. Additionally based on resource availability and costs, the energy source can also be switched to solar thermal energy or electricity-based heating systems.

These technology options are commercially available and with technical progress, their investment cost is declining. Nonetheless, for small firms, individual investment in this new technology may require access to capital that they do not have. An option that could be considered in such cases is a cluster-based approach, whereby a group of similar firms in a location could access a common low-carbon power generating facility (such as a biogas boiler) set up to serve the cluster. Financial support for this would need to be provided initially by either a local or a national body.

Other technologies such as green H_2, carbon capture and storage (CCS) or carbon capture, utilization, and storage (CCUS) are also emerging, which have very high potential to mitigate emissions from the industrial sector. However, as these technologies are still in a nascent stage of development, they are not included in the model and are discussed separately (see Appendix: Emerging Emission Mitigation Options).

The low-carbon scenario assumes that the clean energy transition for industry can be achieved successfully. It projects a reduction of 2.2% in the final energy demand from industry in 2050, as compared with the BAU scenario, primarily due to transition to cleaner and more efficient energy sources, electrification, and the application of energy efficiency practices. The total energy requirements of the industrial sector reach 32.9 Mtoe in 2050, with natural gas meeting roughly two-thirds (21.6 Mtoe). The total specific energy consumption for industrial thermal applications is 23.3 Mtoe, while electric energy consumption is 9.63 Mtoe. This pathway is estimated to mitigate emissions of 363.3 $MtCO_2e$ between 2020 and 2050, across oil, gas, coal, electric, and off-grid-based technologies. The weighted overall marginal cost of abatement is negative and estimated to be $8.54/$tCO_2$e. Gas as a fuel offers the largest potential impact, with a mitigation of 349.86 $MtCO_2$e between 2020 and 2050 and a negative marginal cost of abatement of $ 11.46/tCO_2e.

Table 21 lists the enablers and barriers that must be seized or overcome if the low-carbon strategy is to be achieved for industry. Key will be the pricing strategy, which avoids artificially cheapening environmentally inefficient options and the availability of finance for the transition.

Table 21	Barriers and Enablers for Industrial Energy Efficiency	
Parameters	**Enablers**	**Barriers**
Cost	• Cost savings due to energy conservation.	• Gas is cheap at present so that currently it may not be cost-effective to install energy-efficient boilers.
Policy and Regulatory Environment	• Sustainable Renewable Energy Development Authority (SREDA) Act of 2012 focuses on cutting the energy consumption of energy-intensive industries.	• Suitable regulatory measures and incentive mechanisms are needed for energy savings in the sector. • Insufficient industrial environmental protection measures. • Subsidized energy prices for the sector lead to waste and send the wrong signal.
Market	• 17% of the natural gas supply is used in industrial sectors (Isharat et al. 2015). • Use of old or mal-maintained machines and poor energy management in manufacturing. • Older natural gas boilers can be replaced with high-efficiency or super-high-efficiency units. • Captive generators used by firms are small capacity and are operated inefficiently in open cycle mode.	• Consumption of natural gas has exceeded the discovery of reserves in the last 15 years. • Financing of new technologies is a challenge.
Technical and Commercial Performance	• About 50% of national primary energy is consumed in the industrial sector, so there is huge potential for energy savings.	• Huge investment and time are required for implementation. • Risk government may pivot from coal to imported LNG for power production instead of renewables. • Lack of investment funds is preventing the government from building new plants and shutting the older ones in the fertilizer industry.
Stakeholder Acceptance	• Increasing awareness among firms about their role in sustainable development.	• Delay in response and insufficient resource allocation. • Limited public budget for energy efficiency.
Skill Development Requirement	• Several training and certification courses are organized by SREDA. • Operation and maintenance for natural gas-based heating is like conventional systems.	• Workforce training is required to switch to sustainable methods. • As some of the technologies are still not widely used, a specialized workforce is needed.

Source: Compiled by authors.

4. Investment Cost and Financing Options

Currently per capita energy consumption is low (9.7 gigajoules in 2020) in comparison with both the global average (71.4 gigajoules) and the average of the Asia and Pacific region (59.6 gigajoules). Similarly, the country's per capita electricity generation for 2020 was only 426 kWh and consumption for the same year was only 378 kWh, despite 97% of the population having access to electricity. Currently, renewable sources do not contribute significantly to generation and a major expansion of these is critical to meet both rising demand and emission reduction goals. The low-carbon scenario addresses this by assuming a rapid shift to the low-carbon pathway. This section discusses the investment requirements for different sectors and power technologies rising from the model and the possible funding sources for the substantial investment needed for a clean energy transition. These numbers depict the incremental cost for the additional installed capacity over the already existing capacity within the system averaged on an annual basis. Requirements under the BAU and low-carbon scenarios are compared.

Additional Investment

Business-as-Usual Scenario

The total investment requirement under the BAU scenario in Bangladesh is estimated for three periods: 2020–2030, 2030–2040, and 2040–2050. The costs are $60 billion (2020–2030), $57.3 billion (2030–2040), and $41.8 billion (2040–2050) in constant 2010 US dollars. Under this scenario, major investments in power would be required for coal supercritical, gas combined cycle and nuclear power generation. Simultaneously, a significant investment is also required for upgrading and expanding the electricity grid. Within demand sectors, major investment needs arise from the residential and transport sectors to support their electrification. By 2050, some limited investment for the development of solar PV grid and off-grid technologies are also included in the BAU scenario.

Table 22 shows the distribution of energy system investment from 2020 to 2050, as 5-year totals.

Table 22	Investment Requirement Under the Business-as-Usual Scenario ($ billion 2010)					
Sectors	2020–2025	2025–2030	2030–2035	2035–2040	2040–2045	2045–2050
Agriculture	0.66	0.85	0.55	0.79	0.72	0.74
Commercial	0.39	0.41	0.58	0.61	0.67	0.71
Cooking	0.51	0.49	0.46	0.36	0.27	0.30
Industry	0.35	0.58	1.30	1.26	1.18	1.09
Residential	0.71	0.99	1.35	1.78	2.29	2.94
Transport (Freight)	2.77	1.96	4.35	4.24	4.79	4.80
Transport (Passenger)	0.26	0.26	0.51	0.67	0.77	0.78
Domestic (Resources)	0.00	0.00	0.03	0.01	0.01	0.01
Gas	0.16	0.21	0.27	0.20	0.13	0.27
Imports	0.29	0.31	0.30	0.27	0.00	0.01
Oil	0.01	0.01	0.02	0.02	0.01	0.01
Elec (Off-grid, Diesel Gen Set)	0.00	0.05	0.00	0.01	0.06	0.02
Hydro Power	0.00	0.03	0.03	0.03	0.05	0.03
Onshore Wind	0.00	0.00	0.00	0.00	0.00	0.00
Power Plant (Biomass)	0.00	0.00	0.00	0.00	0.00	0.00
Power Plant (Coal, Supercritical)	14.23	13.92	5.38	2.29	0.90	0.92
Power Plant (Gas, Combined Cycle)	2.71	4.79	5.73	6.86	5.61	4.33
Power Plant (Nuclear)	4.55	4.39	7.31	6.09	2.03	0.00
Power Plant (Oil, Combined Cycle, Peak)	0.00	0.00	0.00	0.00	0.14	0.80
Solar Photovoltaic	0.03	0.07	0.09	0.04	0.13	0.05
Electricity Grid	1.44	1.54	1.68	1.81	1.90	2.33

Notes: Figures are quinquennial totals. Domestic Resources includes domestic solar off-grid, wind off-grid, coal, gas, oil, and biomass.

Source: Compiled by authors.

Low-Carbon Scenario

Similarly, total investment needs under the low-carbon scenario are estimated to be $61.3 billion (2020–2030), $64.9 billion (2030–2040), and $105.7 billion (2040–2050). In total, over the period, this is 45% higher than in the BAU scenario. In the low-carbon scenario, major investments are expected in solar, wind, and biomass technologies for decarbonization of the energy supply. Another major difference is the much larger investment required in the domestic resource sector under the low-carbon scenario due to increased investment in domestic gas and off-grid solar systems used across agriculture, residential, and industry sectors. As in the BAU scenario, the majority of the demand-side investments are expected in the transport and residential sectors, particularly targeting electrification measures. Table 23 shows the projected investment in the low-carbon scenario, again in 2010 dollars.

While the government is committed to decarbonize the economy, as reflected in its international commitments and other policy statements, it is largely dependent on financial support from various bilateral and multilateral donors, international financing institutions, intergovernmental organizations, development agencies, and foundations. Under the low-carbon pathway, in 2010 constant prices a total investment of $232 billion of additional investment is estimated to be required over the period. This is $73 billion higher than investment needs of the BAU scenario, which provides an alternative less environmentally friendly pathway for growth.

Table 23	Investment Required Under Low-Carbon Scenario ($ billion 2010)					
Sectors	2020–2025	2025–2030	2030–2035	2035–2040	2040–2045	2045–2050
Agriculture	0.64	0.97	0.52	0.94	0.84	1.09
Commercial	0.41	0.41	0.60	0.63	0.69	0.72
Cooking	0.39	0.28	0.24	0.25	0.20	0.31
Industry	0.45	0.64	1.41	1.27	1.77	1.71
Residential	0.71	0.99	1.35	1.78	2.29	2.94
Transport (Freight)	2.77	1.96	4.35	4.24	4.79	4.80
Transport (Passenger)	0.28	0.35	0.63	0.89	1.13	1.32
Domestic (Resources)	0.00	0.90	3.63	5.67	13.65	17.39
Gas	0.23	0.24	0.28	0.16	0.09	0.32
Imports	0.26	0.21	0.18	0.06	0.00	0.00
Oil	0.00	0.01	0.01	0.02	0.00	0.01
Hydro Power	0.14	0.33	0.05	0.05	0.05	0.05
Offshore Wind	1.06	3.77	0.00	0.68	0.78	1.73
Onshore Wind	0.00	1.60	0.55	1.87	0.00	1.14
Power Plant (Biomass)	0.04	4.44	0.86	1.21	1.23	5.62
Power Plant (Coal, Supercritical)	0.00	0.00	0.00	0.00	0.08	0.08
Power Plant (Gas, Combined Cycle)	0.81	2.35	3.82	4.36	5.40	6.70
Power Plant (Nuclear)	4.55	4.51	9.06	6.25	0.00	0.00
Power Plant (Oil, Combined Cycle, Peak)	0.00	0.00	0.00	0.00	0.21	0.23
Solar Photovoltaic	12.00	9.57	0.00	3.23	5.77	16.13
Electricity Grid	1.46	1.59	1.75	2.07	2.23	2.27

Source: Compiled by authors.

As these figures are in constant prices, current price financing requirements will be very much higher. In view of the high financing needs associated with the low-carbon transition, ways must be found of accessing additional finance, including particularly from international sources. The following sections discuss funding options.

Private Investment

To attract private investment appropriate fiscal incentives and investment guarantees, as well as a supportive investment climate will be needed. As of June 2021, private participation in total generation capacity accounted for around 43% with a capacity of 9,481 MW. Out of the total privately owned generation capacity, only 116 MW is from solar, while most of the other plants incorporate gas, furnace oil, and high-speed diesel-based technologies. According to the Bangladesh Independent Power Producers' Association, the private sector has invested about $12 billion over the past 10 years through more than 50 power plants.

To address increasing energy needs, the government has been encouraging private sector participation in the power transmission and distribution segments, as well as generation. There is also an interest from private investors in investing as IPPs and as private partners through the PPP initiative. Figure 39 gives investment in electricity through the Private Participation in Infrastructure (PPI) initiative.

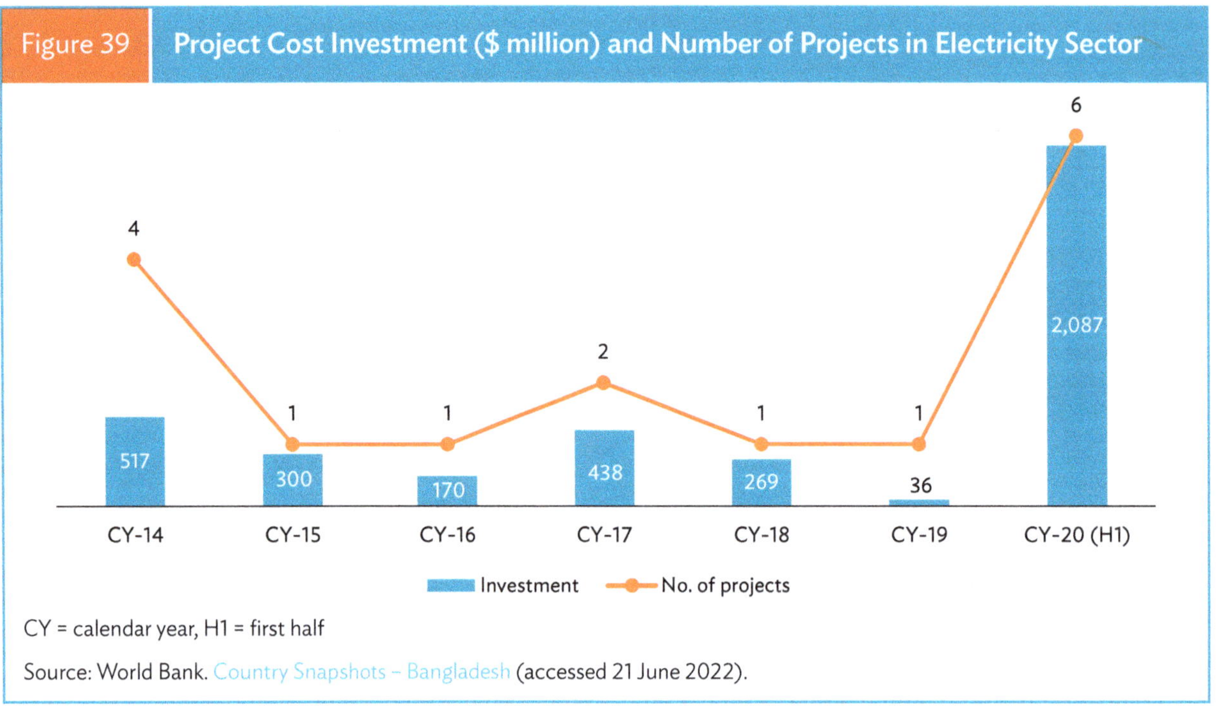

CY = calendar year, H1 = first half

Source: World Bank. Country Snapshots – Bangladesh (accessed 21 June 2022).

Sixteen PPI projects were undertaken over 2014–2020 with the maximum number reported for 2020. This has been attributed to government support through investment incentives like tax exemption on capital gains from transfer of shares by foreign investors, exemption of corporate income tax for a period of 15 years, avoidance of double taxation for foreign investors through bilateral agreements, and no restrictions on work permits for project-related foreign nationals. In 2020–2022, investment in power of about $2,700 million was made in the form of PPI projects. It is working with international institutions, such as ADB, to strengthen the public sector contracting agencies undertaking PPP projects to enhance their capability in attracting private sector involvement in PPPs. However, despite the government's positive outlook toward the involvement of private participants in the energy sector, especially in renewable energy, it requires further support from development partners, multinational financial institutions, and foreign investors if the low-carbon scenario is to be implemented.

Foreign Investment

The capital markets in the country have limited the number of listed securities and therefore have few investment opportunities for portfolio investors. The government, however, is paving the way for the development of the secondary market by introducing derivative products and Alternative Investment Funds. At the end of June 2021, the foreign portfolio stock position amounted to $5.0 billion, while investments in equity securities amounted to $3.3 billion. Investments in fuel and power of $17.78 million accounted for only 0.5% of the total foreign portfolio investments.

FDI is more significant. The government allows 100% foreign ownership and repatriation of invested foreign capital and reinvested dividends in almost all sectors. However, four sectors are reserved for government investments, one of them being production of nuclear energy. Seventeen other sectors are controlled by the government, which include generation, supply, and distribution of power, and these need prior approvals from the relevant government authorities. As of fiscal year (FY) 2020–2021, the total net FDI inflows stood at $2.5 billion, with the power sector receiving net FDI inflows of $456.6 million, accounting for 18.2% of the total net FDI inflows, Figure 40.

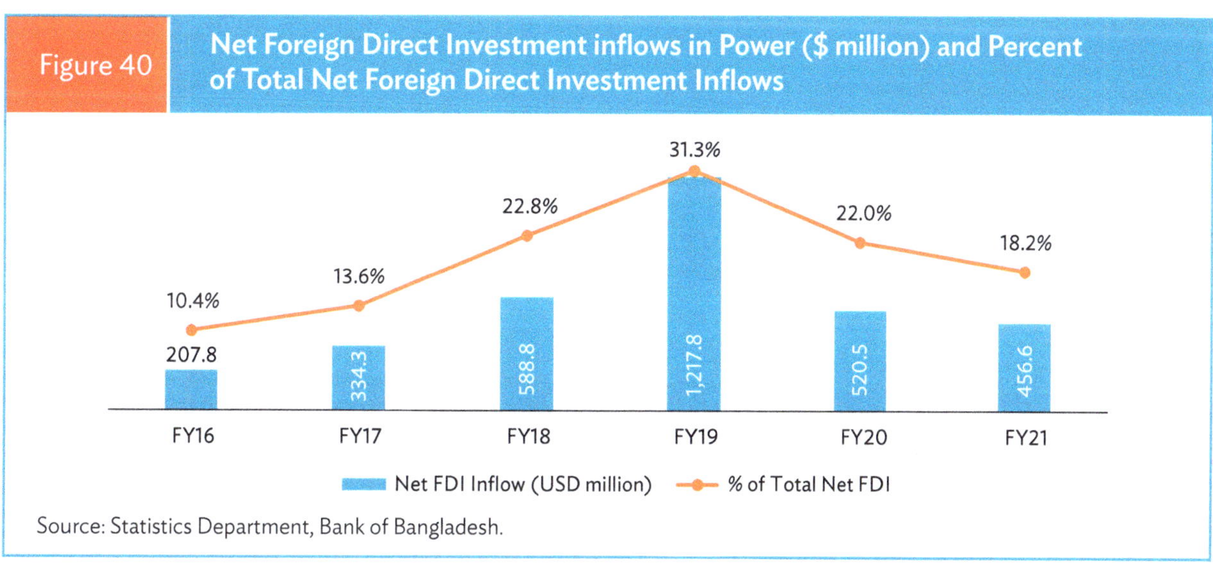

| Figure 40 | Net Foreign Direct Investment inflows in Power ($ million) and Percent of Total Net Foreign Direct Investment Inflows |

Source: Statistics Department, Bank of Bangladesh.

As part of its annual sector specific incentives, the Bangladesh Investments Development Authority has provided the following fiscal incentives to the power and renewable energy sectors:

- Tax exemption of income from private power generation units for 15 years from the date of commercial operations.
- Reduced tax for 10 years for gas pipelines, LNG terminal and transmission lines, solar energy plants, and windmills.
- Import duty exemption on imports of machinery for electricity generation and natural gas exploration.
- No VAT on generation of electricity.

Two examples of FDI power projects are in Table 24.

Table 24	Examples Foreign Direct Investment Power Projects
Project	Details
Rangunia 55-megawatts Electric (MWe) AC Grid-tied Solar Power Plant Project	• Purpose: Develop a solar power plant on a build-own-operate basis • Region: Rangunia, Chattogram • Date of investment: January 2020 • Investor: Consortium of Metito, Jinko Power and AlJomaih Energy and Water (AEW) • Origin of investor: United Arab Emirates, People's Republic of China, and Saudi Arabia
Reliance Bangladesh LNG & Power Ltd	• Purpose: 750 MW LNG-based combined cycle power project • Project cost: Loan of $ 642 million • Region: Meghnaghat, near Dhaka • Date of investment: January 2021 • Investor: Reliance India • Origin of investor: India

Source: Compiled by authors.

Despite having a liberal investment regime, the government has struggled to attract FDI to the country, which is evident from the low net FDI inflows to GDP ratio of 1.29% in FY2018–2019, which reduced to 0.73% in FY2019–2020 and further to 0.71% in FY2020–2021. FDI inflows to the power sector in 2021 had not recovered to 2019 levels, when they were 31% of the national inflow (Figure 41). Failure to attract significant FDI in renewables would be a major set-back to a low-carbon transition.

Bank Credit

Government-owned infrastructure-focused nonbank financial institutions, such as the Bangladesh Infrastructure Finance Fund Ltd and IDCOL, are responsible for the provision of low-cost debt financing to the power sector. Some examples of IDCOL financing are Summit Meghnaghat Power Company Limited (plant: 305–335 MW dual fuel), Energypac Confidence Power Ventures Chittagong Limited (plant: 108 MW heavy fuel oil), and Regent Energy and Power Limited (plant: 108 MW gas-fired). The outstanding total bank credit in 2020 (excluding foreign bills and interbank items) stood at $ 127.63 billion.

As per the Sustainable Finance Policy for Banks and Financial Institutions published in December 2020, there are 11 sectors classified as green and thus eligible for sustainable linked finance. These cover renewable energy; energy and resource efficiency; alternative energy; liquid waste management; solid waste management; recycling and manufacturing of recyclable goods; environment-friendly brick production; green or environment-friendly establishments; green agriculture; green cottage, micro, small-, and medium-sized enterprises; and green socially responsible finance.

The Guidelines for Green Banking have been issued by the government to assess the environmental and commercial risks associated with this lending. These are intended to guide banks to support green policies and incorporate climate risk into corporate risk management practices. The list of active green banking policies is summarized in Table 25. The Central Bank has provided $200 million in a Green Transformation Fund (GTF) through concessional credit to commercial banks, who lend green. As of April 2020, the GTF US dollar fund had only disbursed $41 million in seven projects.

Table 25	Key Policies for Green Banking
Circular Number	**Description**
Bangladesh Regulation and Policy Department (BRPD) Circular No. 07/2012	• Bangladesh Regulation and Policy Department issued circular No. 07 in 2012 to introduce a unified format to report green banking activities of banks.
Green Banking and Corporate Social Responsibility Department (GBCSRD) Circular No. 04/2015	• GBCSRD issued circular No. 04 in 2015 to direct banks and financial institutions to create a climate risk fund and allocate at least 10% of the corporate social responsibility (CSR) budget to the fund. This funding can be done either through grants or through low-interest financing.
Sustainable Finance Department (SFD) Circular No. 02/2016, SFD Circular No. 02/2017 and SFD Circular No. 02/2020	• SFD issued circular No. 02/2016 directing banks and financial institutions to form Sustainable Finance Units and Sustainable Finance Committees, abolishing green banking and CSR units. • In 2017, guidelines on Environmental and Social Risks Management were issued for banks and financial institutions through SFD circular No. 02. It also issued an excel-based risk rating model to evaluate Environmental and Social Risks in the process of Credit Risk Management. • In 2020, SFD issued Circular No. 02 on Refinance Scheme for Environment-Friendly Products/Initiatives/Projects gives detail of sectors eligible for green financing.
SFD Circular No. 01/2018 and SFD Circular No. 01/2019	• SFD issued circular No. 01 in 2018, giving a new uniform reporting format of quarterly review reports on green banking activities for banks and financial institutions to monitor green banking policy and other regulations. The circular also aims to ensure the quality and uniformity of data provided by banks and financial institutions. • To increase the scope of green finance, in 2019, SDF issued circular No. 01, to approve all scheduled banks and institutions' investment in impact funds under the Bangladesh Securities and Exchange Commission (BSEC) (Alternative Investment) Rules, 2015 and for environment-friendly sectors (this is a broad category that includes funding to support protection for women and children's rights).
SFD Circular No. 05/2020	• GBCSRD circular No. 04/2014 replaced by the SFD circular No. 05/2020 states that from September 2020, the minimum target of direct green finance for all banks and financial institutions is 5% of total loan financing.
SFD Circular No. 01/2021	• To develop green finance, SFD issued circular No. 01 in 2021 on Target and Achievement of Sustainable Finance and Green Finance. It requires all banks and financial institutions to ensure that 2% to 15% of their loans meet broader sustainable financing requirements and to set targets at the beginning of each year.

Source: Compiled by authors.

With these policies and initiatives, the direct green finance disbursement increased from $ 390.1m in FY 2015 to Tk111 billion ($1.31 billion) in FY 2020 at an interest rate of 7–8% (Figure 41).

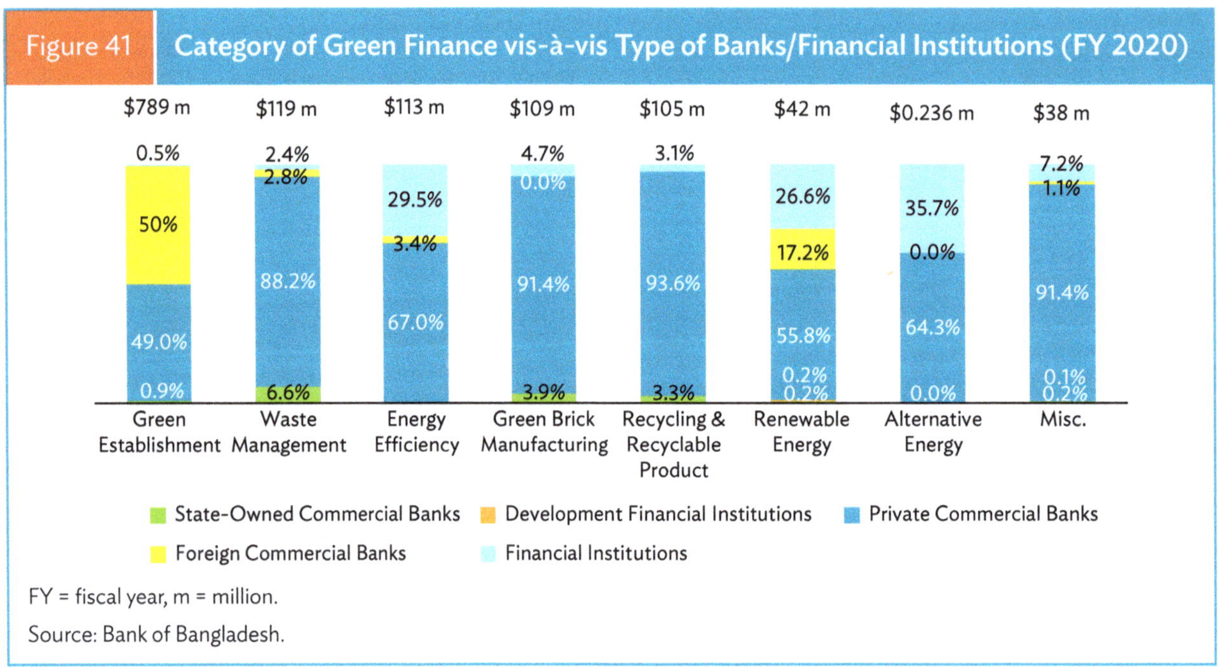

Figure 41   Category of Green Finance vis-à-vis Type of Banks/Financial Institutions (FY 2020)

FY = fiscal year, m = million.

Source: Bank of Bangladesh.

Bonds

The bond market represents 12% of GDP and consists mostly of government bonds and a few corporate bonds. The long-term aim is to develop the bond market to reduce the dependence of private investors on the banking system to finance large-scale renewable projects, especially solar PV projects. However, the market remains relatively small and as of November 2019, the outstanding amount of corporate bonds was only $300 million when compared to $17.2 billion worth of outstanding government bonds. Table 26 gives two examples of bond issuance in the energy sector.

Table 26   Bonds Issued in the Energy Sector

Issuer	Details
Bangladesh-based Beximco Group[a]	• Amount: $0.40 billion in Islamic bonds (sukuk green bonds) • Issue date: Post March 2021 • Description: Bangladesh-based Beximco Group plans to finance two solar PV power plants of 230MW in the northern districts of the country. – Teesta Solar Limited and Korotoa Solar Limited, two subsidiaries of Beximco Power Company Ltd, will take charge of these two projects installing 200 MW and 30 MW, respectively, alongside Chinese partners, TBEA Xinjiang Sunoasis Co. Limited, and Jiangsu Zhongtian Technology Co. Limited
Ashuganj Power Station Company Limited (APSCL)[b]	• Amount: $58.29 face value of corporate bond • Issue date: 2019 • Time for maturity: 7 years • Description: The corporate bond was issued for the construction, commissioning, and operation of the 400 MW (East) CCPP Power Plant project of the company.

[a] S. Islam. 2021. Bond market to finance 230 MW solar projects in Bangladesh. *PV Magazine*. 5 March.

[b] Non-Convertible and Fully Redeemable Coupon Bearing Bond of Ashuganj Power Station Company Limited.

Source: Compiled by authors.

Green bonds specifically aim to finance climate and environment-friendly projects for sustainability. In March 2021, BSEC under the guidance of the Finance Ministry issued guidelines subsection (1) of section 33 of the Securities and Exchange Ordinance, 1969 (Ordinance No. XVII of 1969), which govern the issue of compliant green bonds in the country. The projects eligible for green bond investment are:

- Renewable energy
- Energy efficiency
- Environmentally sustainable management of natural resources and land use
- Clean or environmentally friendly transportation
- Sustainable water and wastewater management
- Climate change adaptation
- Inclusive green buildings
- Eco-efficient products, production, technologies, and processes

The green bond market is new with the first approval for green bonds issued in 2021 via private placement (Table 27). Further, according to the Climate Bonds Initiative database, the Dhaka Stock Exchange does not have a sustainability-related index.

Table 27	Green Bond Issued by the Government
Issuer	**Description**
NGO named SAJIDA Foundation approved by BSEC[a]	**Amount:** $ 12 million green bond **Issue date:** 2021 **Coupon rate:** Green Zero-Coupon Bond **Tenor:** 2 years **Listing:** Private Placement **Use of proceeds:** Renewable energy and environmentally friendly projects **Description:** This is a debut green bond issued to increase microfinance programs as well as to ensure environmental development by investing in agriculture, sanitation, and solar projects, thereby catalyzing climate-friendly initiatives.

[a] Dhaka Stock Exchange; *The Financial Express.* 2021. BSEC approves green bond for first time in Bangladesh. 8 April.

Source: Compiled by authors.

Some of the Central Bank GTF has been invested in green bonds. The IFC, in partnership with the Swedish International Development Cooperation Agency, has undertaken a study to assess the potential for a domestic green bond market in Bangladesh. IFC estimated total climate-friendly investment opportunities of about $172 billion between 2018 and 2030, based on the country's INDC target as well as sectoral targets. The green building investment potential from residential and commercial building was particularly large, as confirmed by the low-carbon scenario in the present study. The potential market for green bonds, is thus very large, but remains largely undeveloped.

International Financial Assistance

Multilateral and Bilateral Agencies

Multilateral development banks (MDBs) are key stakeholders in energy governance because of their involvement in both national and global policy formulation. MDBs are also committed to supporting climate finance. In 2019, the climate finance of MDBs in South Asian countries was $8.3 billion, with Bangladesh accounting for 25%. Bangladesh's total climate financing from 2015 to 2019 was $5.8 billion, of which the total funding for the energy sector is estimated to be $ 1.1 billion. This is the sum of MDB adaptation and mitigation finance, covering the following activities, energy, transport and other built environment and infrastructure, renewable energy, energy efficiency, lower-carbon and efficient energy generation, low-carbon technologies, and non-energy GHG reductions (Figure 42).

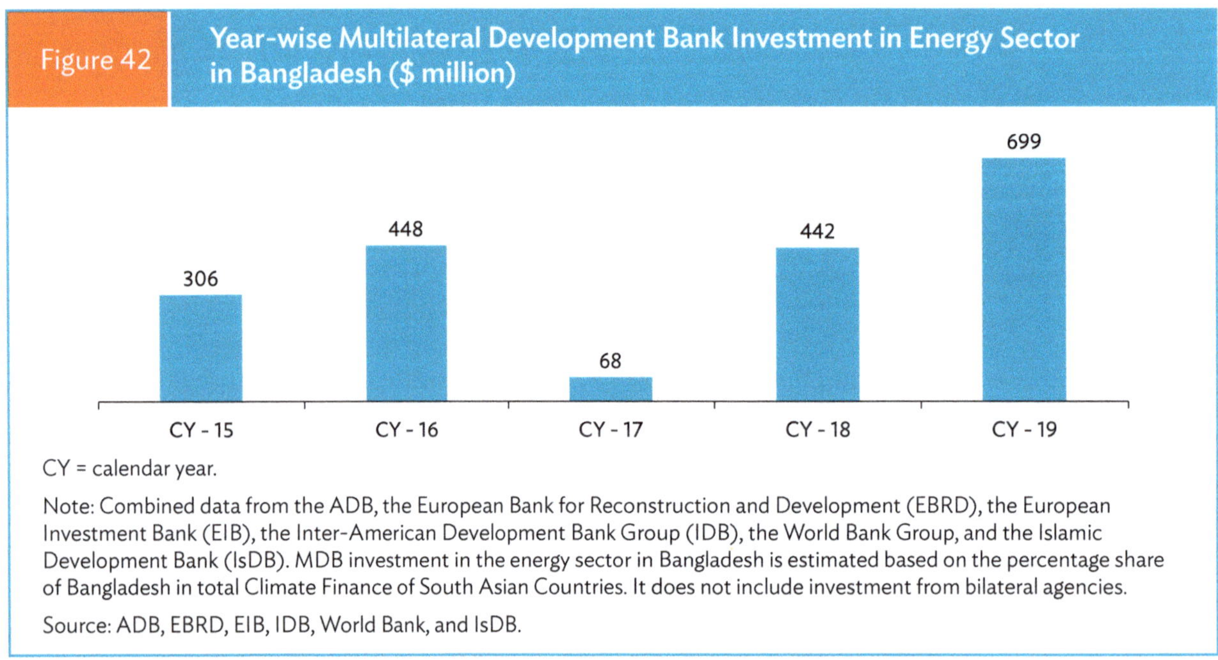

| Figure 42 | Year-wise Multilateral Development Bank Investment in Energy Sector in Bangladesh ($ million) |

CY = calendar year.

Note: Combined data from the ADB, the European Bank for Reconstruction and Development (EBRD), the European Investment Bank (EIB), the Inter-American Development Bank Group (IDB), the World Bank Group, and the Islamic Development Bank (IsDB). MDB investment in the energy sector in Bangladesh is estimated based on the percentage share of Bangladesh in total Climate Finance of South Asian Countries. It does not include investment from bilateral agencies.

Source: ADB, EBRD, EIB, IDB, World Bank, and IsDB.

In 2016 the total investment in climate finance was $448 million, which decreased to $68 million in 2017, before rising to $699 million in 2019. These figures may appear large in absolute terms but are relatively modest compared to the investment needs of the low-carbon (and BAU) scenarios. Hence much more needs to be done.

Table 28 gives examples of the support provided by MDBs (World Bank and ADB) and a bilateral agency Japan International Cooperation Agency (JICA) for the development of the energy sector in the country. The multilateral International Fund for Agricultural Development (IFAD) started its operations in Bangladesh in 1978 focusing on poverty alleviation in the rural population and enhancing food security. Increasingly IFAD emphasizes the integration of renewable energy into its business model to accelerate poverty alleviation in rural areas. It uses the Renewable Energy for Smallholder Agriculture (RESA) method to achieve sustainable rural transformation and bridge the rural energy gap, which is a key obstacle to smallholder agricultural production and post-harvest value addition. Two examples of IFAD RESA projects in Bangladesh are in Table 29.

In addition to funds from MDBs, bilateral agencies have also provided support for the clean energy transition. The United States Agency for International Development (USAID), for example, has provided funds through its Scaling Up Renewable Energy (SURE) program aimed at supporting the government in enhancing private sector engagement, energy security, competitive and transparent renewable energy procurement, and institutional capacity to integrate affordable renewable energy into the energy system.

In addition to funds from MDBs, bilateral agencies have also provided support for the clean energy transition. Table 28 has an example of the support from JICA. As another example, USAID has provided funds through its SURE program aimed at supporting the government in enhancing private sector engagement, energy security, competitive and transparent renewable energy procurement, and institutional capacity to integrate affordable renewable energy into the energy system.

Table 28	Examples of Multilateral and Bilateral Agency Support for the Energy Sector
Agency	**Description**
Scaling-up Renewable Energy Project	• Project detail: It aims at market expansion in renewable energy sector through establishment of utility-scale solar PV and rooftop PV with an energy generation capacity of 310 MW. The project will also aid the IDCOL for the management of the Renewable Energy Financing Facility for both rooftop and utility-scale solar PV. It also aims to support SREDA, find sites for large-scale projects, and promote a new net metering policy for rooftop PV. • Funding Agency: World Bank • Total Investment amount: $185 million • Approving year: 2019
Bangladesh Power System Enhancement and Efficiency Improvement Project	• Project detail: The project aims to provide efficient and reliable electricity supply to over 150,000 rural households in southwestern Khulna division by 2030. • Funding Agency: ADB • Total Investment amount: $200 million • Approving year: 2017
Matarbari Ultra Super Critical Coal-Fired Power Project (V)	• Project detail: The project will construct an ultra-supercritical coal-fired power plant with an output of 1,200 MW in the Matarbari area of Cox's Bazar District, Chattogram Division, in southeastern Bangladesh. • Funding Agency: JICA • Total Investment amount: ~ ¥143,127 million ($1,240.57 million) • Approving year: 2019
Energy Efficiency and Conservation Promotion Financing Project (Phase 2)	• Project detail: The project will facilitate installation of energy efficiency and conservation (EE&C) equipment in Bangladesh and will support promotion of policy in EE&C by extending concessional loans and other support. • Funding Agency: JICA • Total Investment amount: ~ ¥20,076 million ($174.01 million) • Approving year: 2019

Source: Compiled by authors.

Table 29	Examples of International Fund for Agricultural Development Projects in Bangladesh
Projects	**Renewable Energy Activities**
Promoting Agricultural Commercialization and Enterprises (PACE)	• Approval date: 17 September 2014 • Duration: 2014–2022 • Total project cost: $129.79 million • IFAD financing: $58.07 million • Description: Aims to introduce technology and improve farm management practices for small producers and micro-enterprises. For example, during the progress of the project, solar fish dryers were introduced for value chain development.
Climate Adaptation and Livelihood Production (CALIP)	• Approval date: 15 September 2011 • Duration: 2011–2022 • Total Project cost: $133.31 million • IFAD financing: $71.14 million • Description: The program introduced renewable energy technologies such as flexi-biogas systems, bamboo briquettes, and solar home systems. The flexi-biogas system harvests combustible gases from everyday biomass (such as kitchen waste and manure). Benefits of this program planned to include increased fish production and fishing income, increased local employment, increased income from diversified production, and investment in new areas, and low-cost village protection work.

Source: Compiled by authors.

Clean Development Mechanism

Clean Development Mechanism (CDM) offers a potential funding source for low-income countries who can offset emissions; however, Bangladesh's CDM potential is low because the country's main source of energy is natural gas. There is a modest scope for CDM projects, such as detection and repair of natural gas leaks from transmission lines, and programs for solar home systems, improved cookstoves, and energy-efficient brick kilns. As of 1 September 2021, Bangladesh had 21 active registered projects with the UNFCCC as CDM projects, out of which 10 are registered as project activity and 11 are registered as program of activities. Out of the total 21 projects, one is for renewable energy and 10 are energy efficiency projects. As of 1 September 2021, 57% of all registered projects had reached the issuance stage, with a total certified emissions reduction of 13.3 million tons of CO_2, of which 35% is from energy efficiency projects. Realistically future funding for the energy transition under this heading will be limited.

Novel Ways of Funding for Energy Efficiency

Bangladesh has a history of donation and fundraising methods, which are like crowdfunding. For example, in 2012, the government funded the Padma Bridge project by inviting all citizens to make donations of any amount to designated bank accounts. Projekt.co, the country's first crowdfunding platform, was launched in 2015 as reward-based crowdfunding, a type of financing in which entrepreneurs solicit donations from individuals in exchange for products or services, and in 2018 it evolved into equity crowdfunding and peer to peer (P2P) lending.

The government is yet to formulate a policy and regulatory framework for equity crowdfunding and P2P lending. None the less there have been examples in the energy sector (Table 30).

Table 30	Examples of Crowdfunding in the Energy Sector
Crowdfunding Platform	**Details**
SOLshare supported by IIX Impact Partners (Crowdfunding Platform)	• **Description:** SOLshare installed the world's first cyber-physical P2P solar shared power grid in remote areas of Bangladesh. It enables people to buy more electricity at any time and invest in more electricity generation. • **Funding:** Part of funding is supported by IIX Impact Partners (debt and equity crowdfunding platform) through its investor network. • **Achievement:** About 30 P2P solar power trading networks have been installed in Bangladesh and India.
FUND SME	• **About FUND SME:** It is a platform that allows entrepreneurs to share their business plans with a network of investors, venture capitalists, and business supporters. • **Project:** e-Trycatch Technologies Ltd listed a project related to kitchen waste used for biogas production, which provides an effective solution for waste management. • **Project cost:** Tk2 million ($0.023 million) • **Waste requirement:** 1-ton waste everyday • **Revenue per month:** Tk66,000 ($769.95) • **O&M cost per month:** Tk41,000 ($478.30) • **Life of project:** 10 years

Source: Compiled by authors.

Monetizing Energy Savings

The EECMP was launched in 2016, in accordance with Energy Efficiency and Conservation Rules, 2014. The plan establishes energy efficiency and conservation (EE&C) policies, programs, legal documents, and frameworks. Various targets have been set and programs launched to support them (Table 31).

Table 31	Programs for Energy Efficiency and Conservation
Policies/Program	**Details**
Energy Management Program for Large Industrial Energy Consumers	• Designated large energy consumers such as textile and garments, chemical fertilizer, steel making, and rerolling producers are required to reduce total energy consumption through energy audits and energy efficiency improvements. Sustainable Renewable Energy Development Authority (SREDA) enforces and follows up on energy consumption of the designated consumer. • The EE&C potential in Bangladesh is estimated to be 21% of the energy consumption of the entire industry sector. • Under the SREDA Act, 113 energy-intensive industries have been identified as designated consumers.
Energy Efficiency Labeling Program	• The energy efficiency labeling program aims to promote the sale of high-efficiency products in industrial, commercial, agricultural, and residential markets. • Energy-savings potential is about 28.8% in the residential sector, 21% in industry, 20% in agriculture, and 10% in commercial.
Energy Efficiency Building Program Based on the BNBC	• This is an updated version of the BNBC and is a core plan to promote energy-efficient buildings. It includes reducing heat transfer from the outside through heat insulation, air-tight doors or windows, and sunlight control, energy-efficient building equipment and electrical appliances, and the operation and maintenance of buildings and construction equipment.
Energy Efficiency and Conservation Finance Program for Private Companies	• EE&C finance program provides finance to promote innovative pilot projects to strengthen renewable energy and energy efficiency projects. The finance is given as low-interest loans, subsidies, and preferential taxes and is available at an interest rate of 8% for the residential sector and 4% for industry. • The government has launched an energy efficiency financing project to raise awareness in industries. The program was funded by JICA at an interest rate of 4%.
Green Purchase Program for Eco-Friendly Public Procurement and Certification Programs	• The government initiatives include the "Green Purchase Program for Eco-Friendly Public Procurement" and ISO14001 and 50001 certifications programs, applying to local government, state-owned companies, quasi-government organizations, and other parts of the public sector.
Energy Consumption Data Collections	• The program supports the national energy management program. The program is designed to monitor the country's energy consumption data, by fuel, sector, and subsector. • The Ministry of Power, Energy and Mineral Resources provides energy supply data for policymaking. • SREDA provides energy demand information for decision-making and consumer awareness and to help analyze energy consumption, including energy intensity. • National Statistics Bureau receives data on energy sales and production from energy suppliers and industry associations.
Global Warming Countermeasures	• The measures include the creation and quantification of a national carbon market, capacity building in the form of a carbon abatement project, and awareness-raising.

[a] SREDA and Power Division. 2015. Energy Efficiency and Conservation Master Plan up to 2030, accessed on 20 July 2023.

Source: Compiled by authors.

Energy savings can provide funding for further energy-efficient activities. Energy-savings companies (ESCOs) offer a model that has been used in other countries. They provide comprehensive energy-saving services, such as energy-saving solutions, equipment installation, maintenance, and operation. The payment source for this type of ESCO service comes from the energy savings achieved, and the payment amount will not exceed the total amount of the customer's current energy bill. There are two main ESCO business models, the guaranteed savings model, and the shared savings model. Under the guaranteed savings model, the customer obtains funds from the bank based on credit and repays the loan at the saved energy cost. Under the shared savings model, the ESCO provides performance guarantees as well as financing. The energy saved is allocated between the customer and the ESCOs based on a predetermined rate. There are also quasi-ESCOs in which no one provides performance guarantees, but financial institutions agree to provide financing based on the expected cash flow of their energy-saving projects.

While not yet developed in the country a World Bank report on *Demand-Side Energy Efficiency Opportunities in Bangladesh* argues that the ESCO model can play an important role in energy audits and the design of cogeneration systems, low-cost financing for energy-saving chillers, implementation of energy efficiency in the ready-made garment industry and in setting up a clean production center for the textile-dyeing industry. An example of a pilot ESCO project from Bangladesh is in Table 32.

Table 32	An Example of an Energy-Saving Company Project
ESCO Project	**Details**
A pilot project under ESCO model to demonstrate the energy efficiency potential of LED tubes[a]	• **ESCO:** With the support of the government, German Aid (GIZ) implemented the Renewable Energy and Energy Efficiency Program with Sustainable Renewable Energy Development Authority as executing agency. This is a pilot project in the garment sector under the ESCO model to demonstrate the efficiency potential of LED tubes and the feasibility of ESCOs in Bangladesh. • **Term:** 2017–2018 • **Project detail:** GIZ provides financial support for the implementation of the pilot project, and the ESCO supplies and installs standard LEDs in the factory and provides performance guarantees. • **Achievement:** Around 4,400 lights were replaced by LED light, resulting in energy saving of about 1,582 kWh/day. Overall, the project can save about 17.2 MWh of electricity per month and reduce 11.5 tons of CO_2.

[a]Alam. 2017. Paving the Way for Efficient Lighting in Bangladesh's industries, Appropriate Technology. *Burnham.* 44 (1): 57 (accessed on 29 June 2021).

Source: Compiled by authors.

In addition, there has been some support for energy-efficient solar household systems for low-income communities both through funds provided through the CSR activities of banks and companies and microfinance initiatives. All these novel models should be encouraged; however, they are unlikely to provide the large sources of funds needed for the energy transition. These will need to come from a combination of significant external climate finance and private investment both domestic and foreign.

5. Low-Carbon Pathway for Dhaka City

Introduction

Globally, major cities around the world account for around 78% of total energy consumption and over 60% of all GHG emissions.[1] With the rapid rate of urbanization and economic development, these shares are expected to increase further in the coming years. Therefore, it is crucial to study and analyze the energy consumption and emissions profiles of major cities and their role in sustainable energy and climate systems in the future. This section details the energy outlook of Dhaka City and like the national analysis identifies a low-carbon pathway for future economic and sustainable development of the city.

Dhaka is the capital and largest city of Bangladesh. With a population of more than 10 million people, Dhaka is a megacity with two city corporations: Dhaka North and Dhaka South (Population and Housing Census 2022). In this study, we have considered statistics and information available for Dhaka's central region comprising the city corporation areas only. Dhaka City's development has been mostly spontaneous, where the urban infrastructural development has not kept pace with the population growth of the city, raising concerns regarding long-term sustainability. Unplanned urbanization has created an unsustainable pressure on urban services such as housing, transport, water supply, sanitation, and drainage. Further, the city is facing adverse climate effects because of increased air pollution due to high reliance on fossil fuels in demand sectors such as transport, industry, and power. Transitioning toward cleaner energy sources is critical for the city to align with the national commitments and help deliver the national emission reduction targets. Currently, the city is facing several climate-related challenges with adverse impacts on the health of its residents. The air quality has deteriorated over the last decade due to a rapid change in the vehicular fleet, increased congestion and construction, and a large increase in economic activity, resulting in increasing mortality and morbidity among the residents. Some of the major challenges in the city include its high population density, over-dependence on congested roads, inefficient vehicles, unplanned urbanization, and air pollution.

Although most policies pertaining to the low-carbon transition and sustainable development are at the national level, there are certain policies and pilot projects specifically focused on Dhaka City planning and its future road map. The Rajdhani Unnayan Kartripakkha (RAJUK), which falls under the Ministry of Housing and Public Works, the agency responsible for coordinating urban development in Bangladesh, has prepared the Dhaka Structure Plan (2016–2035). The structure plan provides a strategic direction for the Dhaka Metropolitan Region covering the present status of the city, policy measures on effective land use management, the transport sector, the residential and housing sectors, the upgrading of public facilities, and other aspects of governance and sustainable institutional development.

---

[1]   United Nations Climate Action. Generating power.

The federal government has undertaken various initiatives toward provincial development and low-carbon energy transition in the city in recent years. National Solar Energy Roadmap for 2021–2041 identifies Dhaka as one of the provinces with significant potential for solar PV pump-based irrigation. As per the Power System Master Plan (2016), several policy measures for improvement of traffic conditions in urban areas have been considered. As power demand density of the city is estimated to increase further in the coming years, the government has also planned for upgrading the existing power network and power stations in and around Dhaka City. Also, it is actively focusing on the use of LNG technology for the Floating Storage Regasification Unit to bridge the existing supply and demand gap in the city region, particularly for power and industry.

Methodology and Assumptions

A detailed bottom-up analysis had been carried out using the same MESSAGEix model as in the national study to develop a low-carbon scenario for the city. The objective of the study is to identify the key areas for low-carbon development, technology interventions and their impact on the energy demand, and overall emissions of the city energy system. Based on an energy system analysis, the model estimates the investment requirement of the various technologies over 2020-50. As in the national study a BAU and a low-carbon scenario are contrasted. The definitions are as before

- *Business-as-usual scenario*—assumes continuation of past and current trends in the future. This scenario derives future patterns of technological penetration and resource consumption based on historical trends.
- *Low-carbon scenario*—assesses options for achieving energy and emissions reduction trajectories based on the assumption of enhanced ambition for low-carbon development to support Paris Agreement goals.

Modeling provides a useful platform for analyzing the impact of various mitigation pathways and associated costs given specific assumptions around socioeconomic growth, the costs, performance, and availability of technologies, as well as the timing, strength, and scale of mitigation policies. However, availability of reliable good quality data is important for developing a robust model and conducting an accurate modeling exercise. Despite the importance of the subject, data availability at a city level is often a constraint, as in the case of Dhaka City in particular, for industry. Most of the data published at regular intervals is available at the district or division level. Therefore, in places assumptions have had to be used to overcome the data gaps. For the city-level assessment, the energy demand and supply data base incorporated in the model was developed through a series of virtual consultation seminars conducted with local experts and consultancies.

A possible limitation of this study is that waste-to-energy, a renewable form of energy relevant in the city's context, has not been considered. Another study suggests that potential energy ranging from 200 to 2,200 GWh of electricity could be generated from municipal waste in Dhaka City by 2050 under different scenarios (Islam 2016). However, the omission of waste-to-energy in the model is unlikely to be important for the overall results, as this energy estimate is small in comparison with the total electricity needs of the city. In addition, stringent efforts, policy and regulatory measures, and incentives around municipal solid waste management would be needed for the success of this approach. This, however, does not undermine the need to explore the waste-to-energy process for tackling solid waste issues in the city, while simultaneously fulfilling the energy demand of residents.

As in the national analysis the BAU scenario projections are based primarily on the current and historic utilization patterns of energy commodities such as coal, oil, gas, biomass, and hydro. The uptake of emerging renewables and cleaner technologies such as solar, wind, and nuclear is guided by cost. The low-carbon scenario, on the other hand, takes into consideration a set of policy and technology measures to steer the economy toward a low-carbon path. These measures include demand-side electrification, supply and demand-side efficiency improvement, and fuel-switching across supply and demand sectors toward carbon-neutral or low-carbon fuels, such as natural gas, biofuels, and off-grid renewable energy generation. The INDC targets are not explicitly provided as an input rather the scenario outcomes are evaluated to investigate the potential of such interventions to help in achieving the INDC targets.

As the model is based on the city area it has a less comprehensive coverage than the national study. The city model omits agriculture and includes only road passenger and freight traffic. Population-based indexing is used to estimate passenger kms and end-use demand. Electricity is assumed to be imported from the national grid.

The list of low-carbon interventions used in the low-carbon scenario is in Table 33.

Table 33	Summary of Low-Carbon Interventions Adopted in the Low-Carbon Scenario for Dhaka
Scenario	Target Description
Fuel-shift	By 2050 almost 99% of residential cooking would be met through electric cooking technology like induction cookers.
	More than 30% electric vehicle penetration in the road transport sector by 2050.
	Use of natural gas in industries to be replaced by electricity.
Renewable energy	About 20% of the total energy demand in residential buildings (excluding cooking) would be met by solar rooftop PV by 2050.
	More than 25% of the total electricity demand in commercial buildings would be met by solar rooftop PV by 2050.
	By 2050, 3% of total energy demand in industry would be met by solar off-grid.

Source: Compiled by authors.

Total Final Energy Demand Dhaka

Gas and electricity are the main sources of in energy 2020 (Figure 43). In terms of users, the largest share in demand is for cooking, followed by industrial use (Figure 44).

Figure 43	Share of Fuels in the Energy Demand of Dhaka in 2020

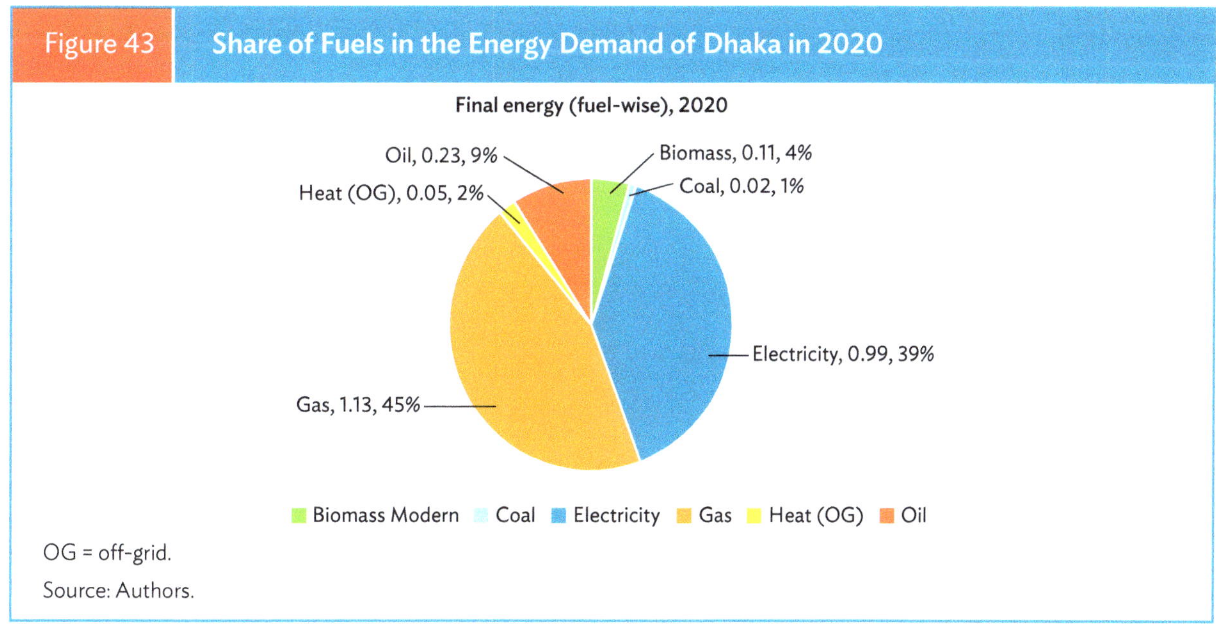

OG = off-grid.
Source: Authors.

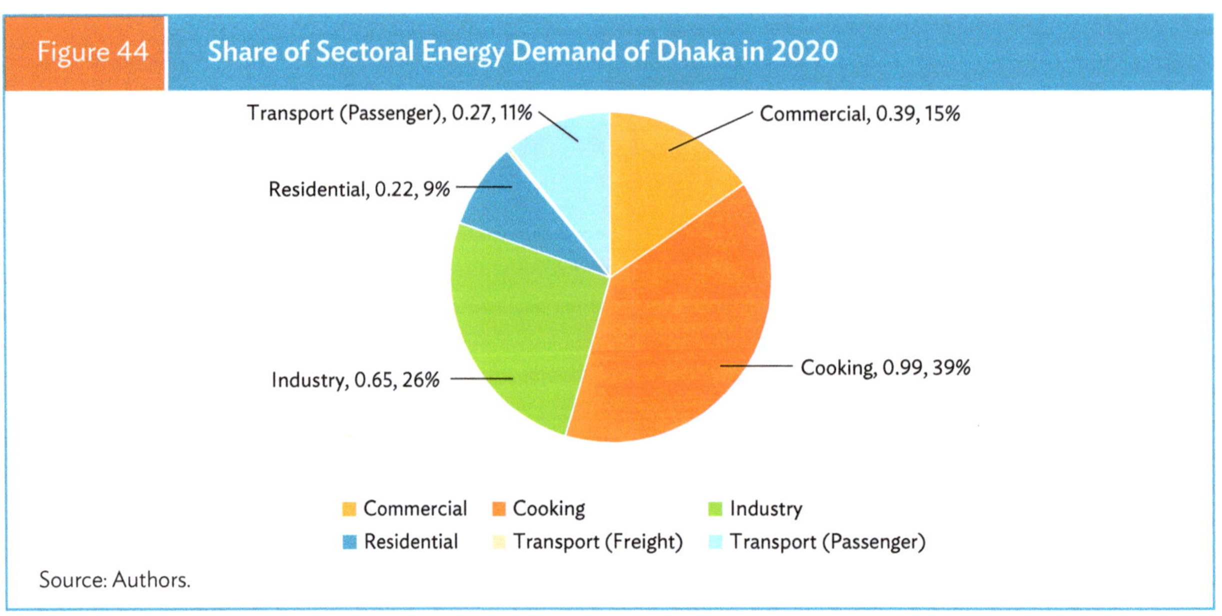

Figure 44   Share of Sectoral Energy Demand of Dhaka in 2020

Source: Authors.

Under the BAU scenario, TFED is estimated to grow from 2.53 Mtoe in 2020 to approximately 9.62 Mtoe in 2050. In contrast, under the low-carbon scenario with measures for efficient utilization of energy, demand is predicted to be 8.81 Mtoe by 2050, a reduction of 8.4% as compared with the BAU scenario. Out of the total energy demand reduction between the two scenarios in 2050 of 0.81 Mtoe, a reduction of 0.97 Mtoe would be from industry, followed by 0.55 Mtoe from cooking. An increase of 0.49 Mtoe and 0.23 Mtoe is projected in the commercial and transport (Passenger) sectors, respectively, due to higher urbanization and increased vehicles in the city. Figure 45 gives the sector-wise energy demand reduction in the low-carbon scenario compared with the BAU in 2050.

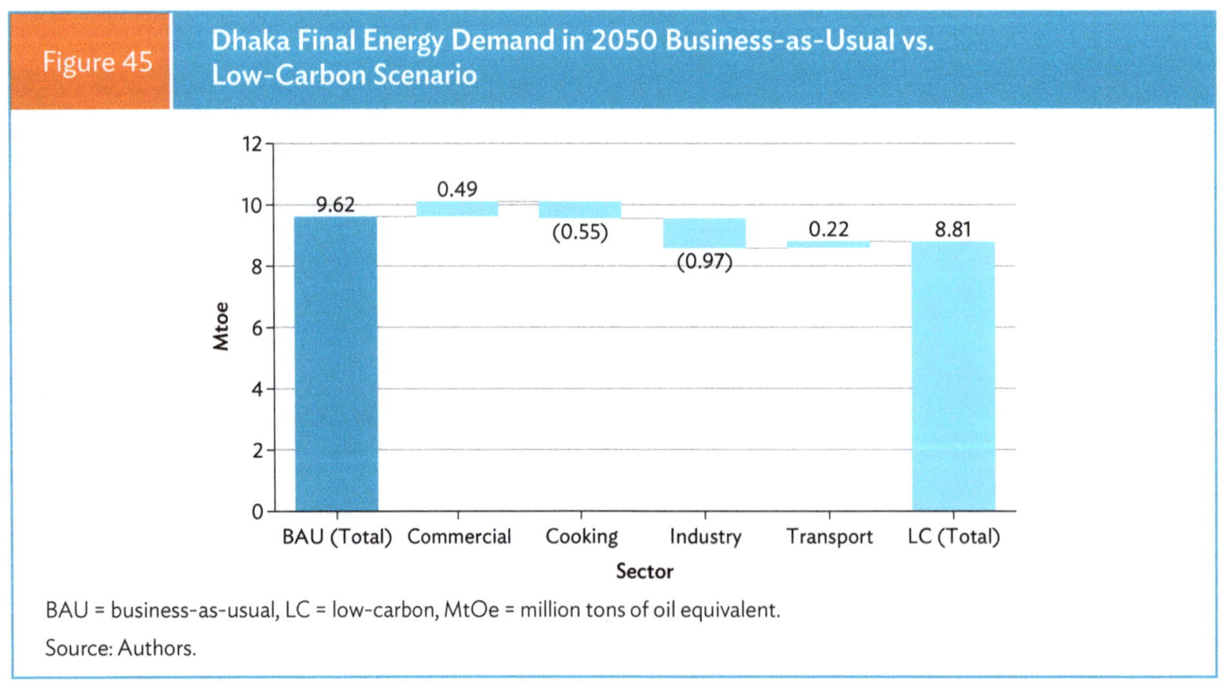

Figure 45   Dhaka Final Energy Demand in 2050 Business-as-Usual vs. Low-Carbon Scenario

BAU = business-as-usual, LC = low-carbon, MtOe = million tons of oil equivalent.
Source: Authors.

The reduction in final energy demand will also influence the primary energy requirement of the city. Over the years, the TPES is estimated to grow from 2.77 Mtoe in 2020 to 10.76 Mtoe in 2050 under the BAU scenario. Under the low-carbon scenario, it is estimated to be 5.7% less at 10.15 Mtoe in 2050. While the absolute primary energy supply remains relatively similar, there is a significant change in the energy supply mix in the low-carbon scenario with a rise in the share of solar PV and in lower-carbon electricity imports from the national grid, and a fall in the share of gas. The combined share of solar PV and solar thermal increases from 3.4% in 2020 to 11.9% in 2050 under the low-carbon scenario, indicating a shift toward cleaner energy sources, predominantly solar PV off-grid, applications across the residential, commercial, and industry sectors (Figure 46).

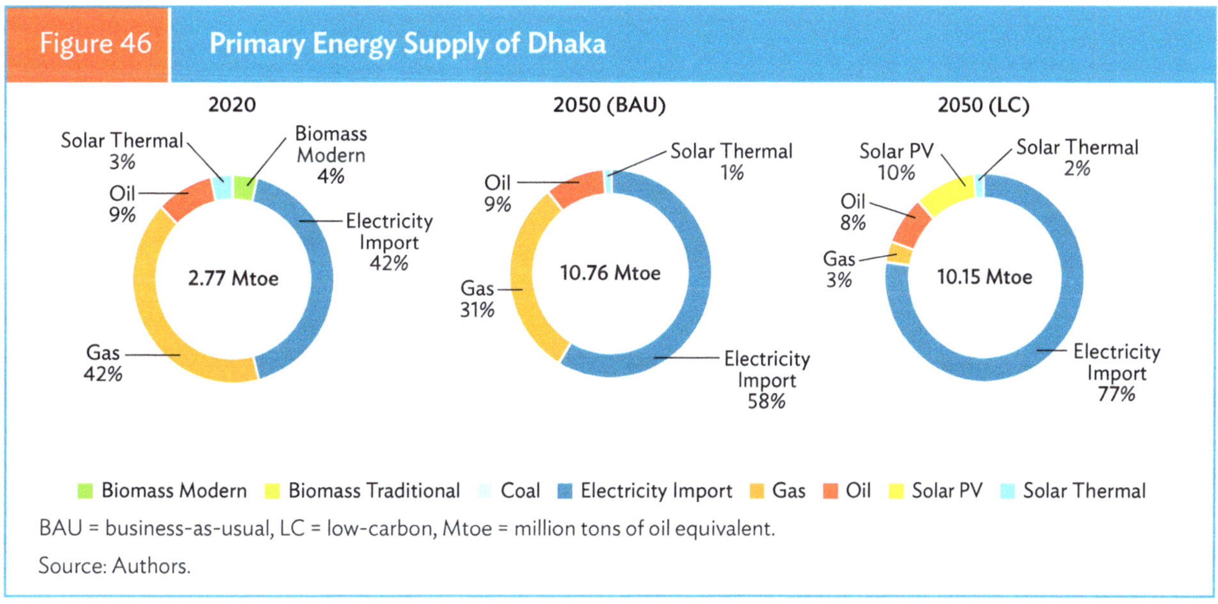

Figure 46   Primary Energy Supply of Dhaka

BAU = business-as-usual, LC = low-carbon, Mtoe = million tons of oil equivalent.

Source: Authors.

Electricity Supply for Dhaka

At present, Dhaka is almost entirely dependent on imported power from power stations located outside the city. Only a very small amount of power consumed in the city is produced there from solar off-grid applications. Under the BAU scenario total electricity generation for consumption in Dhaka grows at a CAGR of 5.8% from 2020 to 2050. Electricity generation will be 13.5 TWh in 2020, 42.6 TWh in 2040, and 73 TWh in 2050 and the vast majority of electricity will be supplied from the national grid (accounted here as electricity imports) with solar PV expected to remain a minor share, although some of the imported electricity could be solar based.

Under the low-carbon scenario, grid-connected solar PV is expected to remain a small share of the total electricity mix due to land constraints in the city hindering the development of solar parks; however, the quantum of off-grid solar installations is considerably higher in the low-carbon scenario than the BAU scenario. Solar off-grid installed capacity is expected to rise from around 29.9 MW in 2020 to 181 MW by 2030 to reach 6844 MW by 2050. Here also, it must be noted that the imported electricity might also be from renewable energy. Figure 47 gives the projected trends in installed solar grid-connected capacity and solar off-grid capacity under the low-carbon scenario.

| Figure 47 | Installed Electricity Capacity by Source for Dhaka Under the Low-Carbon Scenario (2020–2050) |

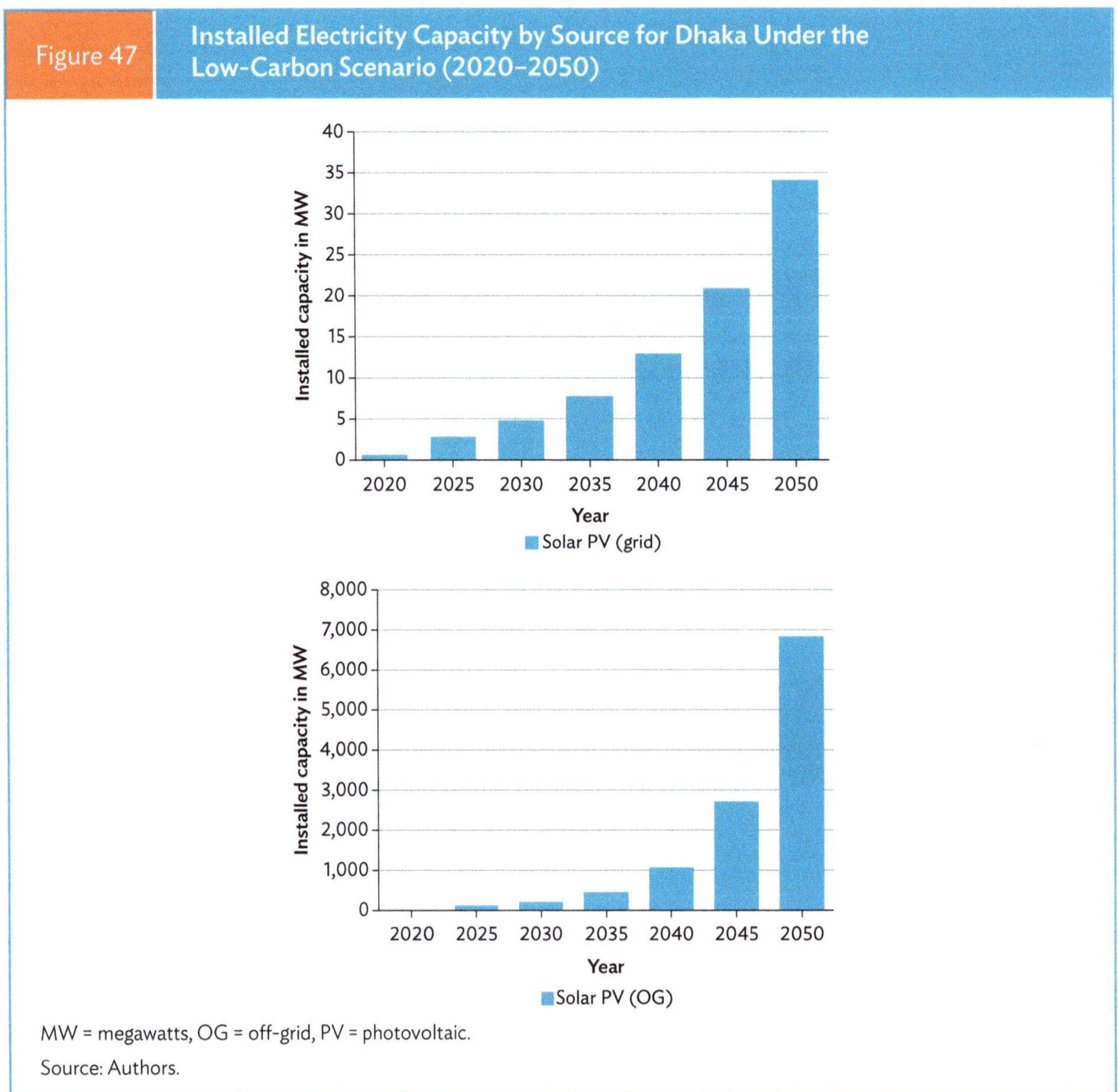

MW = megawatts, OG = off-grid, PV = photovoltaic.

Source: Authors.

Emissions for Dhaka

The model considers emissions factors associated with the use of various energy commodities, which are then used to arrive at future sectoral emissions. In the BAU scenario continued reliance on fossil fuels in demand sectors is the major driver for the growth in GHG emissions. The total GHG emissions (from the supply side and demand side combined) in Dhaka under the BAU scenario grow from 13.5 $MtCO_2e$ in 2020 to 19.2 $MtCO_2e$ in 2030 and 37.8 $MtCO_2e$ in 2050, an annual emissions growth of 3.5%. The transport sector is the largest emitter throughout the period and produces almost 80% of emissions from the demand side from 2020 to 2050 (Figure 48).

Under the BAU emissions from supply side are much lower but are projected to grow at 4.2%. The total GHG emissions from the supply-side increase from 0.13 $MtCO_2e$ in 2020 to 0.19 $MtCO_2e$/year in 2030, rising to 0.46 $MtCO_2e$/year in 2050 (Figure 49).

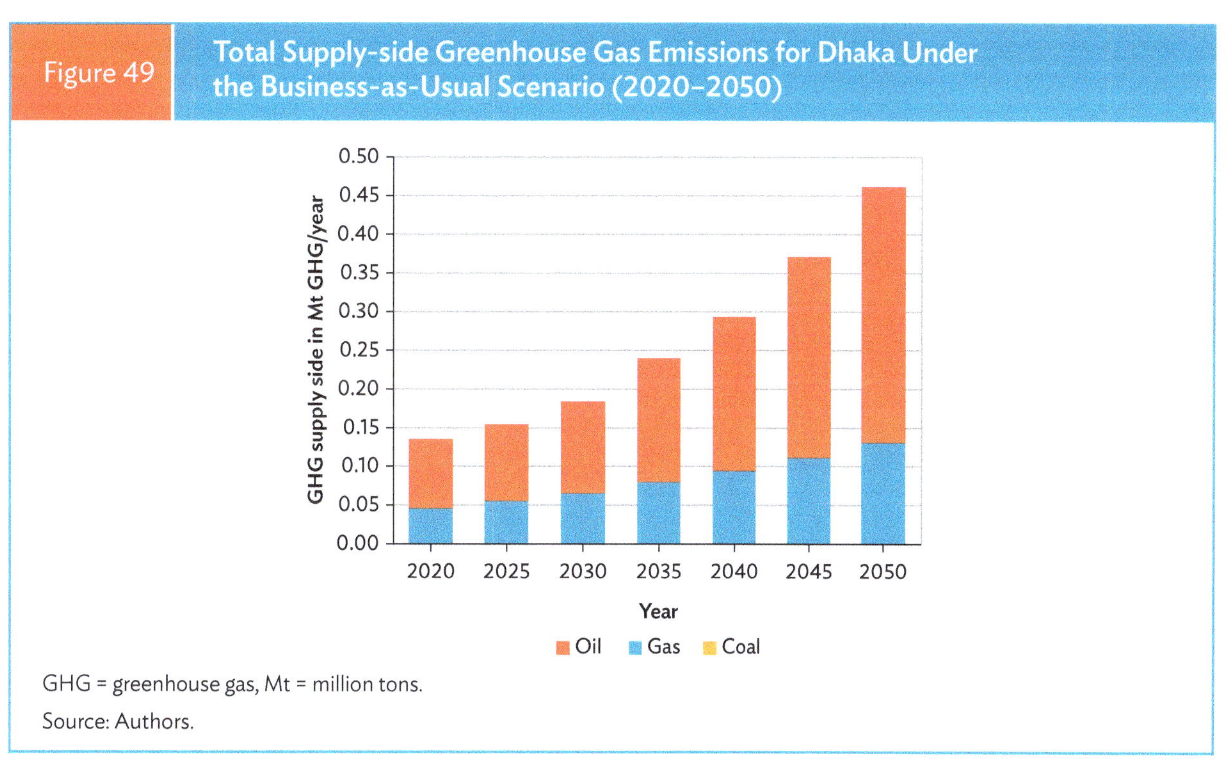

Figure 48: Total Demand-Side Greenhouse Gas Emissions for Dhaka Under the Business-as-Usual Scenario (2020–2050)

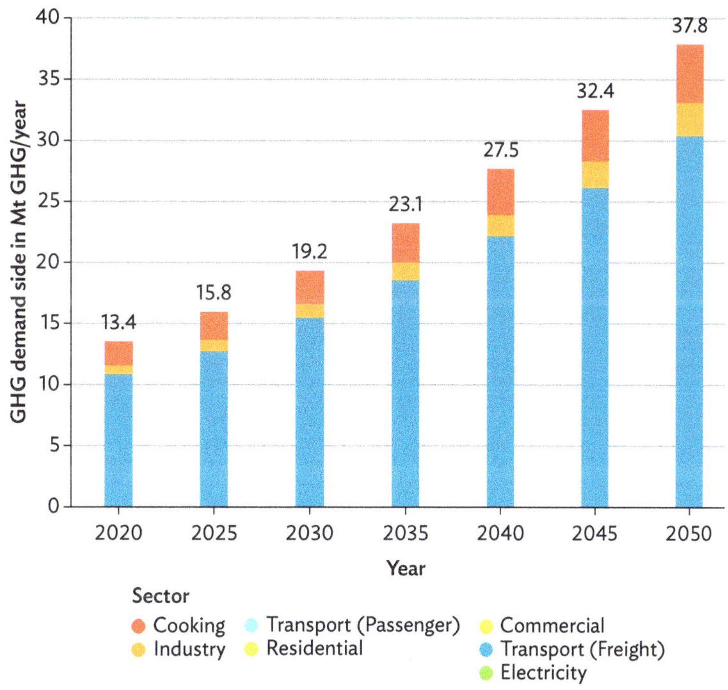

GHG = greenhouse gas, Mt = million tons.

Source: Authors.

Figure 49: Total Supply-side Greenhouse Gas Emissions for Dhaka Under the Business-as-Usual Scenario (2020–2050)

GHG = greenhouse gas, Mt = million tons.

Source: Authors.

Overall, for Dhaka comparing the two scenarios total emissions are 40% lower in the low-carbon scenario, as compared with the BAU scenario. The reduction in emissions is primarily due to energy efficiency interventions in the demand sectors and increasing penetration of clean energy in the overall energy supply mix. The transport and residential sectors, in particular, contribute strongly to emissions reduction in the low-carbon scenario. Transitioning from petrol and diesel-powered vehicles to electric vehicles is estimated to mitigate emissions of 138.8 $MtCO_2$ between 2020 and 2050. The transition from biomass to electric cooking and the greater use of efficient cooking appliances is projected to result in an emission reduction of about 130.7 $MtCO_2$ between 2020 and 2050, of which 120.8 $MtCO_2$ is solely from residential cooking (Figure 50).

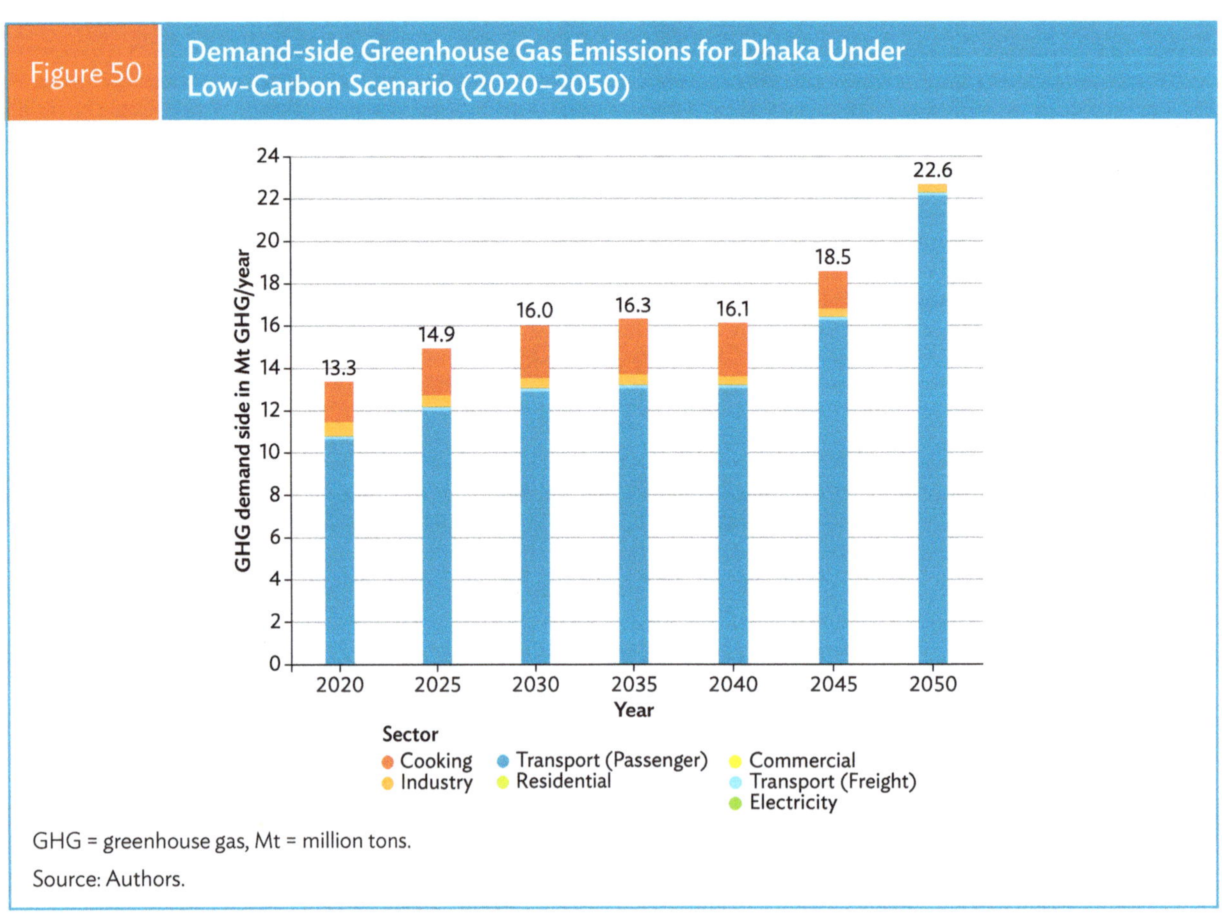

Figure 50 | **Demand-side Greenhouse Gas Emissions for Dhaka Under Low-Carbon Scenario (2020–2050)**

GHG = greenhouse gas, Mt = million tons.
Source: Authors.

The comparison between GHG emissions in the BAU and low-carbon scenarios is shown in Figure 51. In 2050 the low-carbon scenario has 15.4 $MtCO_2$ lower emissions, due almost exclusively to demand-side improvements. In terms of user sectors, saving of emissions are 8.0 $MtCO_2$ in transport, 4.7 $MtCO_2$ in cooking, and 2.5 $MtCO_2$ in industry.

Table 34 gives the estimated total emissions mitigation 2020-50 for Dhaka under the low-carbon scenario. Changes in the residential and transport sectors are the key. Changing cooking from traditional biomass to electricity will increase efficiency and help reduce emissions. Installation of efficient cooling appliances such as inverter ACs and high-efficiency chillers will help reduce total energy demand, while installation of rooftop solar PV will reduce dependence on grid supply and further reduce emissions. In transport the assumed shift from petrol- and diesel-powered vehicles to electric vehicles will create significant reduction in emissions.

Table 35 presents an overview of potential low-carbon technologies applicable for Dhaka.

Figure 51	Dhaka City Emissions in 2050, Business-as-Usual vs. Low-Carbon Scenarios

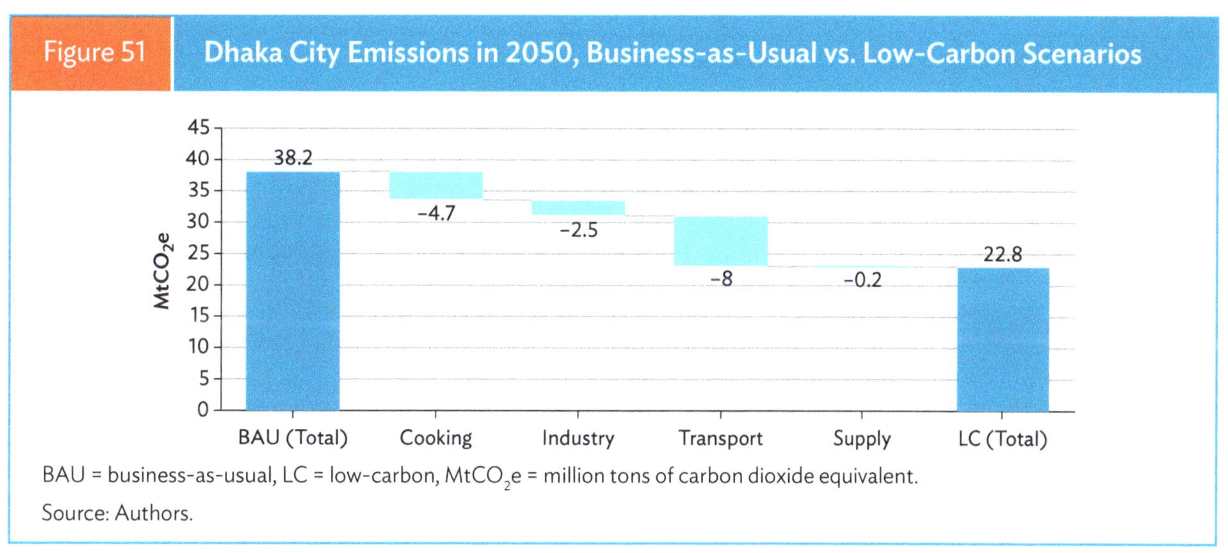

BAU = business-as-usual, LC = low-carbon, MtCO$_2$e = million tons of carbon dioxide equivalent.

Source: Authors.

Table 34	Emissions Mitigation Under Low-Carbon, Dhaka 2020–2050

Source	Emissions Mitigation MtCO$_2$e
Energy Supply	10.8
Residential Sector	130.7
Transport	138.8
Industry	4.81

MTCO$_2$e = million tons of carbon dioxide equivalent.

Source: Compiled by authors.

Table 35	Potential Low-Carbon Technologies for Dhaka City

Technology	Description	Existing Policy & Measures	Country Context
Electric Vehicles (EVs)	Electricity-based technology that would become an alternative to internal combustion engine vehicles.	✓ Shift toward EVs is slower in Bangladesh compared to India and the People's Republic of China	The government has proposed 3.4 million tons CO$_2$ eq. (Mt) GHG unconditional emission reduction from the transport sector by 2030. Automobile Industry Development Policy 2021 targets most new passenger cars, bus, trucks and three-wheeler auto rickshaws will be EVs by 2030. Proposed draft policies under review are Electric Vehicle Registration & Operation Policy 2022 and draft EV Charging Guidelines.
High-Efficiency Residential and Commercial HVAC	Use of inverter technology and fixed speed air-conditioners for better performance along with common metric and standards labeling program to increase savings and quality.	✓ The EECMP has set specific EE&C programs to achieve energy saving in the building sector ✓ The National Housing Policy ✓ BNBC	The stock of air-conditioners in 2019 is estimated at 3.5 million units. Based on the average capacity of room air-conditioners, this translates into 6.0 million tons of refrigeration (TR). Demand is projected to be around 17 TR by 2030.

continued on next page

Table 35	*continued*		
Technology	**Description**	**Existing Policy & Measures**	**Country Context**
Decentralized Generation and Mini-Grids	A mini-grid is defined as a system with a renewable-based electricity generator (with capacity of 10 kW and above), and supplying electricity to a target set of consumers through a public distribution network.	✓ Draft National Solar Energy Roadmap (2020) ✓ Renewable Energy Policy (2015) ✓ Solar Homes Program	It is estimated that Dhaka has about 35,819 thousand square meters of suitable area of rooftops, which have the potential to generate 8.57 terawatt-hours of solar-based electricity.
Solar Photovoltaic Grid	PV panels or arrays are connected to the utility grid through a power inverter unit allowing them to operate in parallel with the utility grid. Extra power generated can be stored in batteries or fed back into the grid.	✓ Draft National Solar Energy Roadmap (2020) ✓ Renewable Energy Policy (2015) ✓ Solar Homes Program	The current installed solar capacity is 416 MW. A 73 MW solar park in Mymensingh started commercial operations in 2021. Plans to set up a solar plant of capacity 100 MW in Jamalpur have been finalized. Under the National Solar Energy Action Plan, the government is aiming for 40 GW of installed solar capacity by 2041. The government also plans to achieve 40% of its power generation with clean energy by 2041.
Mass Rapid Transit (MRT) Options	MRT options such as metro rail, bus rapid transit, and expressways enhance access and reduce congestion.	✓ The Dhaka Metro MRT is a 129-km metro rail system under development.	The Dhaka Transport Coordination Authority developed a 20-year strategic transport plan (STP) in 2005 to develop a more integrated transit system for Dhaka City. The STP was revised in 2016 to reflect present transportation needs considering rapid growth and development in the city. Five MRTs and two bus rapid transit projects are to be developed, according to the revised plan.

Source: Compiled by authors.

Investment Requirements for Dhaka

While the low-carbon scenario sets out a development pathway that lowers the energy requirements and GHG emissions through clean and energy-efficient technologies, as in the national case these benefits are associated with higher investment costs. It is estimated that transitioning toward clean and energy-efficient alternatives in the low-carbon scenario would require additional investments of $28.36 billion (2020 prices) between 2020 and 2050 over the investment requirements in the BAU scenario.

Under the BAU scenario total investment requirements in Dhaka between 2020 and 2030 are estimated to be around $2.03 billion, between 2030 and 2040 $4.27 billion, and between 2040 and 2050 $8.16 billion. Within the demand sectors major investment would be required to serve the residential and transport sectors. Table 36 shows the distribution of energy system investments under the BAU scenario from 2025 to 2050.

Table 36	Investment Requirement Under the Business-as-Usual Scenario for Dhaka ($ million, 2020)					
Sector/Technology	2025	2030	2035	2040	2045	2050
Commercial (Hot Water, Thermal, OG)	2.4	2.4	75.6	5.1	5.1	5.1
Commercial (Other, Elec)	139.2	80.9	196.4	155.9	294.2	280.2
Commercial (Other, Elec, OG)	0	1.6	0	2.3	1.7	3.4
Cooking (Bio)	6.5	5.2	3.8	2.8	2.1	1.5
Cooking (Elec)	4.0	4.1	6.5	13.1	2.9	16.8
Cooking (Gas)	17.9	22.6	28.0	34.5	4.3	5.2
Domestic (Resource—Elec OG Solar Photovoltaic [PV])	17.2	17.9	20.8	62.3	35.7	41.7
Domestic (Resource—Thermal Hot Water OG Solar)	0	0	26.2	0.8	0.8	0.7
Electricity (Grid)	33.9	38.7	50.0	125.0	101.2	136.9
Electricity (Solar PV)	1.4	0.6	1.2	1.8	3.0	4.8
Gas—Transmission and Distribution	7.7	8.8	11.1	40.7	19.6	23.7
Industry	8.8	40.5	14.9	70.2	25.6	32.7
Residential Air-conditioner	88.7	98.8	171.4	212.4	310.6	392.7
Residential (Other Elec)	28.3	43.6	64.9	93.4	130.9	180.3
Residential Hot Water Thermal OG	13.3	0	0	0	13.1	0
Residential Other Elec OG	13.7	18.4	24.4	32.7	41.7	58.9
Transport (Freight)	124.4	124.4	124.4	125.0	125.5	125.5
Two-Wheelers (Elec)	38.1	59.5	95.2	160.1	258.2	416.1
Two-Wheelers (Oil)	235.6	398.7	595.0	940.1	1,409.9	2,082.5
Three-Wheelers (Gas)	129.1	148.8	275.2	404.0	627.7	933.0

Elec = electricity, OG = off-grid, PV = photovoltaic.

Source: Compiled by authors.

In Dhaka under the low-carbon scenario investment requirements are estimated to be around $2.35 billion (2020–2030), $7.61 billion (2030–2040), and $32.85 billion (2040–2050). As noted, the difference from the BAU scenario is additional investment of $28.36 billion (2020 prices) between 2020 and 2050, which is roughly double the investment needs of the BAU scenario. On the supply side, high investments are expected in the solar grid-connected PV and off-grid technologies, while the majority of the demand-side investment is expected to be in the residential sector for efficient appliances and in transport sector for low-carbon options like electric vehicles. Table 37 gives the investment requirements for the city by sector and technology type from 2025 to 2050.

Table 37	Investment Requirement Under Low-Carbon Scenario for Dhaka ($ million, 2020)					
Sector/Technology	2025	2030	2035	2040	2045	2050
Commercial (Hot Water, Thermal OG)	2.4	2.4	71.4	4.3	4.7	6.5
Commercial (Other, Elec)	136.1	80.1	191.6	153.5	287.4	275.4
Commercial (Other, Elec, OG)	3.9	2.6	11.9	22.0	57.1	125.0
Cooking (Bio)	6.5	5.2	3.8	2.8	2.1	1.5
Cooking (Elec)	5.4	7.3	15.2	40.3	30.3	94.0
Cooking (Gas)	18.1	20.8	23.2	23.2	0	0
Domestic (Resource—Elec OG Solar Photovoltaic [PV])	170.2	99.4	372.5	862.8	1,945.7	4,230.5
Domestic (Resource—Thermal Hot Water OG Solar)	0	0	33.3	0	0	0
Electricity (Grid)	35.5	43.7	59.5	142.8	130.9	184.5
Electricity (Solar PV)	1.4	1.0	1.2	1.8	3.0	5.0
Gas—Transmission and Distribution	3.3	2.7	2.1	23.8	0	0
Industry	40.5	59.5	72.6	235.6	152.9	190.4
Residential Air-conditioner, Elec	88.7	98.8	171.4	212.4	310.6	392.7
Residential Other Elec	28.0	42.2	59.5	82.7	99.4	89.3
Residential Hot Water Thermal OG	2.2	2.7	3.6	1.7	3.6	3.6
Residential Other Elec OG	28.6	80.3	220.2	630.7	1,761.2	4,926.6
Transport (Freight)	88.1	77.2	65.5	59.5	53.6	49.4
Two-Wheeler (Elec.)	64.3	173.7	470.6	1,272.5	3,439.1	9,295.1
Two-Wheeler (Oil)	214.0	340.0	508.7	799.7	1,230.5	1,903.4
Three-Wheeler (Gas)	129.1	148.8	273.7	404.0	627.7	933.0

Elec = electricity, OG = off-grid.

Source: Compiled by authors.

Policy Options for Dhaka

Dhaka City is an important part of the clean energy transition. The model suggests that under the low-carbon pathway by 2050 the city will take roughly 10% of national total final demand (8.81 Mtoe out of 115.8 Mtoe) and correspondingly 10% of total national emissions (22.8 $MtCO_2e$ out of 222 $MtCO_2e$). However, its share in demand-side emissions is much higher at around 20% (22.6 $MtCO_2e$ out of 115.8 $MtCO_2e$). This high proportion is likely to be due to the concentration of residential buildings and related activity and the extensive transport links in the city. Its estimated share in emissions mitigation (2020–50) is around 8% (285.1 $MtCO_2e$ out of 3684.3 $MtCO_2e$). Estimated investment requirements are not comparable with the national estimates as they are in different sets of prices.

Advancing the transition in the city will require appropriate national policies (discussed further in Section 6); however, the city administration can play an important role in delivering strong governance and effective implementation of low-carbon programs. A collaborative process for monitoring and evaluating progress could be established consisting of the responsible energy company and city-level officials, relevant local trade organizations, nongovernment organizations, and a Bangladesh Energy Regulatory Commission representative. This body could meet at regular intervals to monitor and assess progress.

It is important to increase public understanding and to generate awareness among city inhabitants about the available low-carbon options and the programs aimed at their uptake. Implementation of efficient air-conditioning systems for district/centralized cooling, for instance, requires the identification of potential buildings that can be developed as a model project to showcase the technology and its benefits, particularly for new areas of the city. The analysis suggests that the key opportunities for the city's decarbonization relate to electrification of the transport, cooking and industry sectors. The interventions in transport and buildings require involvement from both national and local government and the private sector, while power sector interventions, such as to scaling up deployment of solar PV power plants, are mainly driven by national policies.

Electrification of end-use demand also requires simultaneous efforts toward decarbonization of power generation, including additional investment in electricity distribution to meet the increased electricity demand from the electrification of transport and cooking. Solar PV has been suggested as the primary source for decarbonizing the power sector in Dhaka. While large-scale solar PV deployment is in the purview of the government and private sector investors, small-scale decentralized systems can be deployed by individuals in the form of solar PV rooftop systems.

Installment of 7 GW of distributed solar, as envisaged in the low-carbon scenario, is a major task for the city. A host of fiscal, policy, and regulatory measures would be necessary to achieve this scale of transition. The C40 Cities Climate Leadership Group suggests the need for financial incentives such as grants or rebates, subsidized loans, and tax incentives. Regulatory and policy measures could include streamlining the planning and permit issuing process, implementing bulk procurement programs to reduce consumer cost, mandating green building codes, and introducing net metering or feed-in tariffs. Some additional measures include promoting community solar, and public stakeholder involvement to inform program design and to promote distributed solar systems. While various national-level programs support the development of these systems, the city administration can support the governance and implementation of these programs by providing information on the services involved, fast-tracking local government approvals, promoting best practices, and developing demonstration projects.

A city-level low-carbon transition requires close collaboration between the city administration, regional and national governments, and investors to develop innovative and strategic business models targeting the adoption of the measures and investment central to a low-carbon pathway. Access to finance is crucial, so it is important to identify sources of finance and the financial instruments needed, as well as the geographical location of potential investment and the sectors and activities to be supported.

6. Conclusions

This study develops a pathway and a technological roadmap for a clean energy transition in Bangladesh. The government's plans for the transition are reflected in the low-carbon scenario, which makes assumptions about the application of low-carbon options and shows how the country's international commitments on climate targets can be achieved. The roadmap will help policymakers navigate the transition by identifying key technologies and discussing various barriers and enablers to their application. The low-carbon scenario requires higher investment (relative to the BAU case) to achieve climate its goals and the roadmap also identifies the investment needs by sector and technology, as well as discussing potential financial modalities.

Based on the low-carbon scenario, total energy-based emissions are expected to reach around 222 $MtCO_2e$ by 2050, which includes 106 $MtCO_2e$ from supply-side emissions and 116 $MtCO_2e$ from the demand side (Figure 52 and Figure 53). At the UN Climate Change Conference in Glasgow (COP26), subject to the availability of adequate international financial support, the government committed to developing a climate-resilient and low-carbon energy system. The low-carbon scenario shows a pathway for achieving this. The comparison of the BAU and low-carbon scenarios shows that the low-carbon scenario requires very substantial additional investment of $73 billion in 2010 prices between 2020 and 2050.

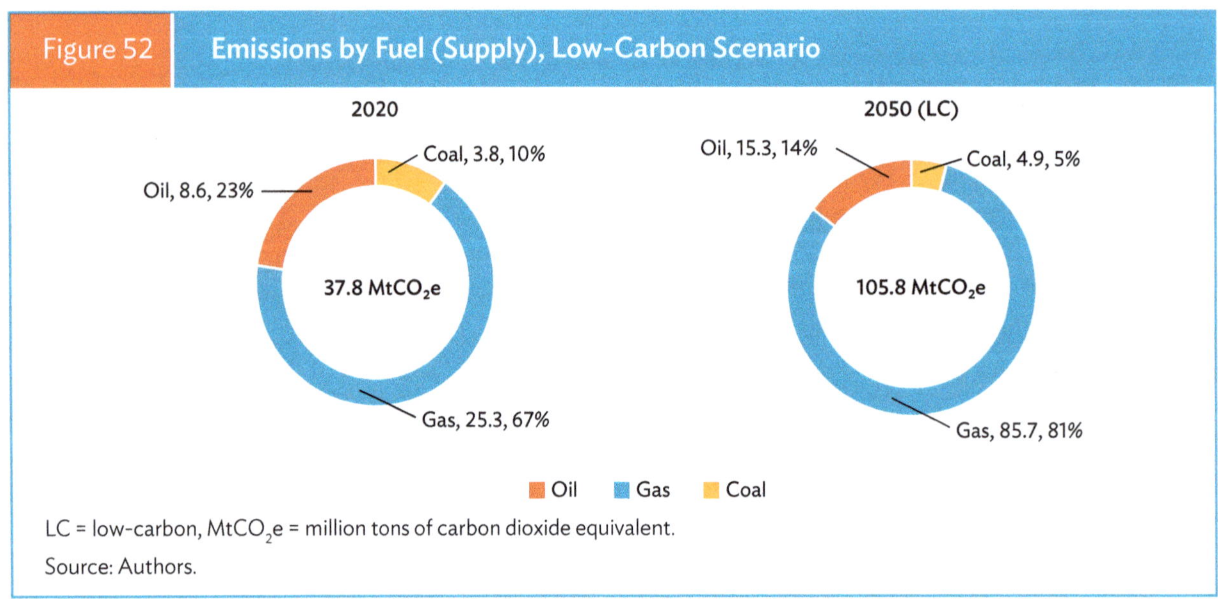

Figure 52   Emissions by Fuel (Supply), Low-Carbon Scenario

LC = low-carbon, $MtCO_2e$ = million tons of carbon dioxide equivalent.
Source: Authors.

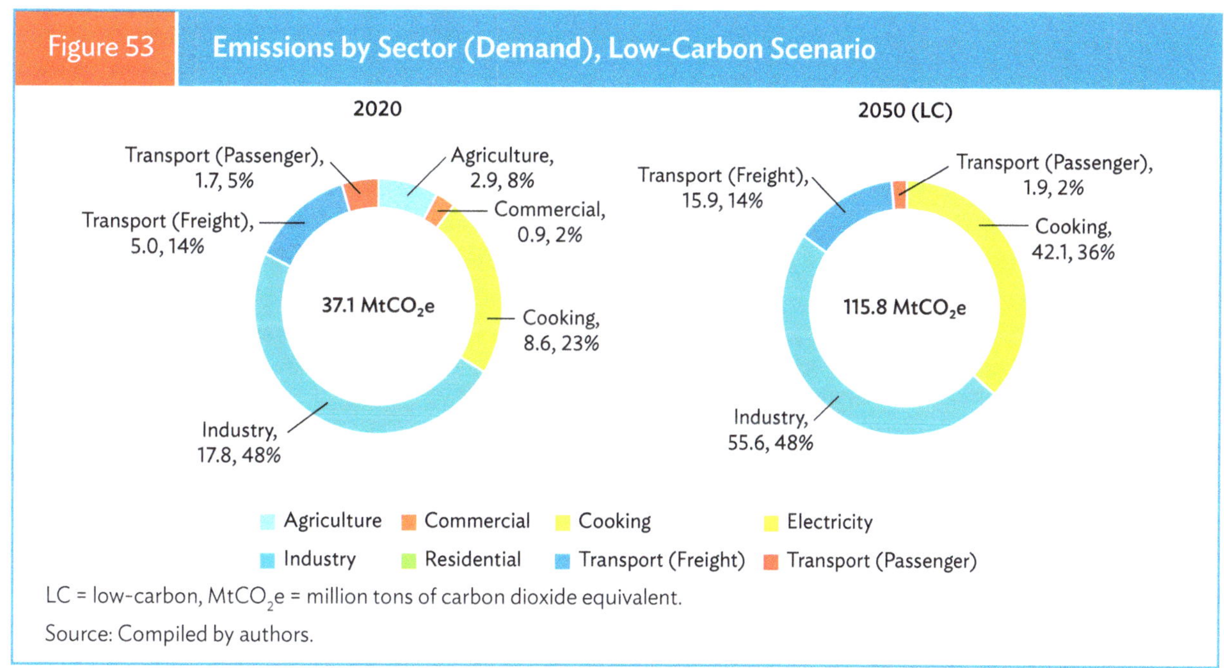

LC = low-carbon, MtCO$_2$e = million tons of carbon dioxide equivalent.

Source: Compiled by authors.

Supply-side Measures

From the low-carbon scenario, a key supply-side measure with high mitigation potential is the switch from fossil-based power generation to renewable-based generation, such as solar PV and wind. Natural gas will play a crucial role in the clean energy transition, as it allows the gradual phase out of coal as a fuel in power generation and industrial production. Deployment of the gas combined cycle power generation technology not only lowers emissions but also enhances the efficiency of resource utilization compared to coal subcritical and supercritical power generation. In the low-carbon scenario total gas-based power generation capacity is estimated to grow from 12.3 GW in 2025 to 23 GW by 2040 and 34 GW by 2050. The combined cycle power plants have a mitigation potential of 716.4 MtCO$_2$e between 2020 and 2050, with a negative abatement cost of –$2.28/tCO$_2$e.

While natural gas-based power generation is comparatively higher cost than coal-based generation in financial terms, it is a cleaner fuel than coal. Moreover, natural gas power plants can support peak demand and provide load following or load balancing services to support the integration of renewables in the grid. Emissions per kWh of electricity generation from gas-fired power plants are thought to be 60–70% lower than comparable coal power plants. While there are many supporting factors for gas-based power development in the country, the high cost of infrastructure development for production and transportation, which has been included in the investment costing in the two scenarios and must be addressed as part of the transition. The low-carbon scenario shows that the demand for natural gas will increase by nearly four times, from 20.1 Mtoe in 2020 to 77.6 Mtoe in 2050. Given this expected trend, the development of existing and the exploration of new natural gas reserves, will be essential, as currently proven reserves are expected to last only for another decade or so.

The low-carbon scenario suggests only limited investment in coal, using the most energy-efficient supercritical coal technology. Under the low-carbon scenario, coal-based power generation will account for 2.4% of electricity generation by 2025, decreasing to 1.3% by 2040 and to 1.1% by 2050, with all generation using supercritical coal technology. Transitioning to an efficient coal-based power generation process is estimated to mitigate 5.1 MtCO$_2$e. between 2020 and 2050.

Since total overall electricity requirements are projected to rise strongly up to 2050 substantial decarbonization of the power sector is required. Under the low-carbon scenario electricity demand is expected to reach 516 TWh in 2050, increasing from 157 TWh in 2025 with a CAGR of 4.8%. The power sector needs to increase clean electricity generation and a significant capacity addition from renewables such as solar-, wind-, and biomass-based power generation is projected, reaching 23% of power generation by 2050.

Investment in solar PV-based power generation is one of key energy transition opportunities identified for the country. Solar applications cover solar home systems, rooftop solar and floating solar, both on-grid and off-grid. There is a high potential for floating solar projects, but a careful study is required since the coastal land and riverbanks in the country are subject to constant erosion and flooding. In the low-carbon scenario, solar PV power generation capacity is estimated to grow from 13.4 GW in 2025 to 48 GW by 2050. The total emissions reduction from solar is estimated as 319.8 $MtCO_2e$ between 2020 and 2050 with an abatement cost of $1.09/$tCO_2$e.

Wind-based power generation has significant untapped potential. With a low installed capacity and immense potential for both onshore and offshore wind farm development, the country has significant scope for wind-based power generation. In the low-carbon scenario, wind power grows from 0.5 GW in 2025 to 10.3 GW by 2050. The total emissions reduction is estimated at 44.3 $MtCO_2e$ between 2020 and 2050, with an abatement cost of $4/$tCO_2$e. Limited domestic manufacturing capacity and construction capabilities for wind farms might be a barrier and there is a need for capacity building and skill development for allow their spread.

Substantial additional financial resources are required for achieving the low-carbon emissions pathway. The growing economy and associated rising electricity demand leads to high investment requirements for additions to power generation capacity, along with the expansion of the transmission and distribution grid infrastructure to deal with renewable energy sources. The model estimates that the capacity expansion in the low-carbon scenario for coal and natural gas-based combined cycle power generation requires $0.16 billion and $23.4 billion, respectively between 2025 and 2050 at 2020 prices. The ambitious levels of additional generation capacity from renewable sources, as envisaged under the low-carbon scenario, require even higher investment. In the low-carbon scenario, cumulative investments between 2025 and 2050 in renewable power capacity additions are estimated to be $13.18 billion (wind), $47 billion (solar), and $13.41 billion (biomass). Considering the significant renewable energy potential in the country and the government's climate commitments, there is an opportunity for domestic firms to produce solar and wind power plant equipment or for international firms to set up local manufacturing units. The government should consider how interest to invest in the clean energy sector can best be stimulated to serve this potentially large market.

In general, on the supply side, supporting the clean energy transition from a fossil-based energy supply will require development of additional infrastructure, expansion of electricity grid capacity, adoption of suitable energy storage operations, and new transmission corridors for power evacuation of green energy. There is also a need to gradually up-skill or re-skill the existing workforce to enable them to adapt to the transition and be a part of the emerging and growing energy sector.

Demand-side Measures

While current practices lack far behind international best practice in relation to energy efficiency, to show what can be achieved the low-carbon scenario adopts strong assumptions on the speed of improvement on the current situation. Key mitigation opportunities on the demand side arise in the transition to gas-based cooking, increasing penetration of efficient appliances in the building sector, the coal-to-gas transition in industry, and electrification of transport. In agriculture under the low-carbon scenario, the transition from diesel to electric and solar-based irrigation pumps is a major aspect of the transition. The deployment of solar irrigation pumps is expected to outpace that of the electric pumps, with a growth rate of 12.6% in the number of solar water pumps deployed between 2020 and 2050. By 2050, the country is expected to have a total of more than 13.4 million water pumps, out of which 88% are projected to be running on solar energy. Electric pumps are also projected to increase at a growth rate of 2.7%, 2020–2050, with a share in total pumps of 11% in 2050, so that diesel-based irrigation pumps are assumed to be largely phased out by this point. The country needs to promote electricity and solar-based irrigation pump sets for sustainable and low-carbon growth in agriculture. Financial incentives may need to be considered to support this shift.

In buildings there is considerable scope for decarbonizing. Enhanced penetration of energy-efficient appliances especially the HVAC system, which is the main user in overall energy demand from buildings, will be important. SREDA has targeted a 20% energy efficiency improvement in buildings by 2030. To achieve this lack of public awareness and regulatory barriers need to be overcome and advanced practices, like construction of net-zero buildings, need to be encouraged.

The low-carbon scenario identifies the transition to electricity- and gas-based cooking as a key emission-reducing measure. The official target, as per CAP for Clean Cookstoves, is to achieve 100% clean cooking by 2030, and this should be encouraged. With the increasing penetration of electric- and gas-based cooking, there will be an additional requirement for gas and a reliable electricity supply, and hence it will be important for these supplies to be provided in as low carbon a way as possible. The low-carbon scenario also identifies the transition away from coal-based generation to natural gas as critical in industry. It predicts a reduction of 2.2% in final energy demand from industry in 2050, as compared with the BAU scenario. This is primarily due to the transition to gas as a cleaner and more efficient energy source, electrification, and the deployment of energy-efficient practices.

Transport is one of the sectors with a high mitigation potential from electrification. Electric vehicle technology is rapidly commercializing, and cost is becoming at par with conventional vehicles. The country, at present, has a draft of a policy to support an electric future in the transport sector, which can include the development of some manufacturing in the country linked with electric vehicles, whether assembly or parts and components. However, it is critically important to develop the necessary electric vehicle infrastructure and ensure the provision of fast charging capacity. While electrification is more likely to be effective for passenger transport, emissions from road-based freight transport can be improved significantly by upgrading norms for emission standards and providing incentives and infrastructure to support their achievement.

Results for Dhaka City

The model has been run separately for Dhaka. Here also in the low-carbon scenario there is a shift in primary energy supply for the city with a rise in the share of solar PV and an increase in the share of cleaner-carbon electricity imports from the national grid. The combined share of solar PV and solar thermal increases from 3.4% in 2020 to 11.9% in 2050 indicating the shift toward cleaner energy sources, predominantly solar PV off-grid applications across the residential, commercial, and industry sectors. Solar off-grid installed capacity is predicted to rise from around 29.9 MW in 2020 to 181 MW by 2030 to reach 6844 MW by 2050.

In terms of efficiency improvements and emissions reduction the city, under the low-carbon scenario changes in the residential and transport sectors are the key. Changing cooking from traditional gas and biomass to electricity will increase efficiency and help reduce emissions. Installation of efficient cooling appliances such as inverter air-conditioners and high-efficiency chillers will help reduce total energy demand, while installation of rooftop solar PV will reduce dependence on grid supply and further reduce emissions. In transport the assumed shift from petrol, and diesel-powered vehicles to electric vehicles will create a significant reduction in emissions.

National Policy Implications

This study highlights the importance of three aspects of the energy transition for achieving a low-carbon growth path-energy efficiency improvement, fuel-switching, and electrification. While the pathway set out in the low-carbon scenario shows what can be achieved and its investment cost, delivery of this pathway will be dependent on the success of policy interventions and access to finance. A strategy to implement a successful low-carbon transition broadly needs to have six dimensions—technological, economic, financial, governance, institutional, and social.

Best available technologies must also be deployed across different sectors. However, local innovation adapting and modifying these to local conditions will be important. Research and development support is crucial for fostering technological innovation and needs to be combined with setting effective standards and certification, to ensure environmental targets are achieved.

Financial incentives and financing are also crucial elements of a low-carbon strategy. Attractive incentives and subsidies can be used to nudge consumer and producer behavior toward cleaner alternatives and efficient use of resources. For this to happen prices in the market need to incorporate environmental costs, so that producers and consumers can adequately take account of the environmental impact of their decisions on what to produce and consume. For example, providing lower rates of road tax or toll charges for electric vehicles or waving city center parking charges for them. A city-level bulk procurement scheme can also be considered by aggregating the demand for energy-efficient technologies and appliances, such as LED bulbs or rooftop solar panels, to lower unit cost for users. Examples of demand-aggregation and bulk procurement models are the Indian LED lighting and US rooftop solar PV schemes.

 The role of financing is critical and various funding sources are discussed in Section 4. Realistically, external finance on a substantial scale, both concessional aid funds and climate finance and foreign investment, will be critical. Involvement of the private sector requires that the investment climate is supportive and the incentive package on offer is competitive.

A strong legal and regulatory framework is required to support interventions, such as the mandatory enforcement of MEPS in residential and commercial buildings or emissions levels from trucks and cars. This must incorporate stringent monitoring and evaluation of progress on achievement of the targets central to a successful low-carbon transition.

Policies and targets alone will not be sufficient without high standard and coherent institutions to ensure their monitoring and enforcement. Institutional coordination is important; for example, the energy and climate ministries can identify and resolve differences between sectoral priorities and policies, and jointly support actions across sectors and institutions. Likewise, national and city-level institutions should coordinate on implementing the necessary policy interventions; for example, the monitoring, inspection, and enforcement of the BNBC could be embedded in the mandates of city corporations.

Finally, public support and engagement will be needed. Social engagement in the form of stakeholder involvement and participation is also a critical requirements for successful decarbonization. It is necessary to increase public understanding and generate awareness about the available low-carbon options and the policies and programs aimed at their uptake. Behavioral change, like a preference for nonmotorized transport and a modal shift toward public transport, can lead to reduction in energy demand. However, consumers must be convinced of the merits of such changes and be made aware of their consequences in terms of environmental impact.

This road map shows how policy change can accelerate a low-carbon transition for Bangladesh. The low-carbon scenario sets out the pattern of development required in the energy sector to create a low-carbon pathway that is compatible with the country's international climate commitments. This is contrasted with the pattern under a BAU scenario, where current trends are continued. While both scenarios allow economic growth, the low-carbon path creates significantly less emissions. It comes with a considerably higher investment cost, and it will be a major challenge to find the funding for this. By identifying the measures required on both the supply and demand sides, as well as the policy change needed to accelerate decarbonization, the roadmap is intended to guide governments, investors, producers, and other stakeholders in relation to investment and production needs and to inform long-term energy planning.

Appendix: Emerging Emission Mitigation Options

This appendix discusses the technologies and emissions mitigation options, which are either currently not cost-efficient or not commercially available, and which therefore are not part of the low-carbon scenario, but which have the potential for future application in the country.

Hydrogen (H_2)

H_2 is primarily used in ammonia production and in the oil refining process. The use of H_2 as a fuel in activities like steel production, transport, and H_2 blending with natural gas is becoming more common globally. Given the current H_2 production process, almost all the H_2 being produced through steam methane reforming or natural gas reforming, is considered as "gray hydrogen." On the other hand, producing H_2 from biomass or electricity by the process of electrolysis or biomass gasification or fermentation is considered as "green hydrogen," as the electricity is from renewable sources. At present, various processes are commercially available for producing both gray and green H_2. Alongside this, several other processes with an emphasis on producing green H_2 are under development. Based on a peer reviewed comparative analysis, out of 10 green H_2 production processes reviewed, only 3 are commercially available at present (biomass gasification, alkaline electrolysis, polymer electrolyte membrane electrolysis), 1 is expected to be available commercially in the near future (solid oxide electrolysis cells), while 6 are expected to take a significant time to become commercially viable (photolysis, dark fermentation, photo fermentation, microbial electrolysis cells, thermochemical water splitting, photo electrolysis) (Comparative review of hydrogen production technologies for nuclear hybrid energy systems).

At present, the H_2 production market is dominated by gray H_2 due to its lower cost of production. As per the United Kingdom H_2 strategy, the levelized cost of H_2 production using the steam methane reformation (SMR) route is one-third of the cost of producing green H_2 using grid electrolysis. Current projections suggest that by 2050 the cost of H_2 production through renewable electrolysis will reach par with that from the SMR process. How far this catch-up can be accelerated is unclear.

In January 2021, the Bangladesh Council of Scientific and Industrial Research (BCSIR) with a pilot processing plant were established with the aim of diversifying the country's energy mix. BCSIR has started training scientists to advance H_2 technology. This should support the future transition to H_2 and develop the market for H_2-based applications. Current research suggests Bangladesh has a potential for green H_2 use in several hard-to-abate sectors, where green H_2 can be used to replace conventional fossil fuels. The following are several possibilities:

- **Hydrogen blended natural gas supply.** H_2 blended CNG is a mix of H_2 and CNG where the share of H_2 can go up to 30%.
- **Green hydrogen** can be used for ammonia production and oil refining, replacing gray H_2 to reduce emissions. For steel a H_2-based direct reduction process can be used to make green steel. The technology is still under development with a TRL 5 and an expectation of commercial availability by 2030. H_2 can also be blended in cement production to minimize emissions.

- **Hydrogen fuel cell (HFC) vehicles.** While electrification can be used to mitigate emissions from light vehicles, there is limited scope for its application in heavy-duty commercial vehicles and shipping. An HFC vehicle can be a potential solution for heavy-duty long-distance freight. Currently the technology is evolving, but growth in the HFC vehicle segment is expected in the future.
- **Hydrogen-based energy storage.** While the use of renewable energy is increasing globally, the issue of grid balancing and the share of variable renewable energy in the grid must be addressed. Electrolysis-based H_2 production is a means of building grid stability and ensuring the maximum utilization of renewable energy. Here, the surplus electricity from the renewable sources is used to produce H_2 to avoid fluctuations.

Carbon Capture Utilization and Storage (CCUS)

Over time, CCUS is expected to become an integral part of the energy system in most countries. Broadly the captured carbon can be transformed in two routes: storage (CCS) and utilization (CCU). Storage means locking in emissions in geological sites for the long term, while utilization means cleaning the captured CO_2 and modifying it for further use, such as carbonated beverages, dry ice, and enhanced oil recovery from oil wells. Broadly, the carbon capture value chain involves the capturing and cleaning of emissions and their transport to the storage or processing facility (Figure A1).

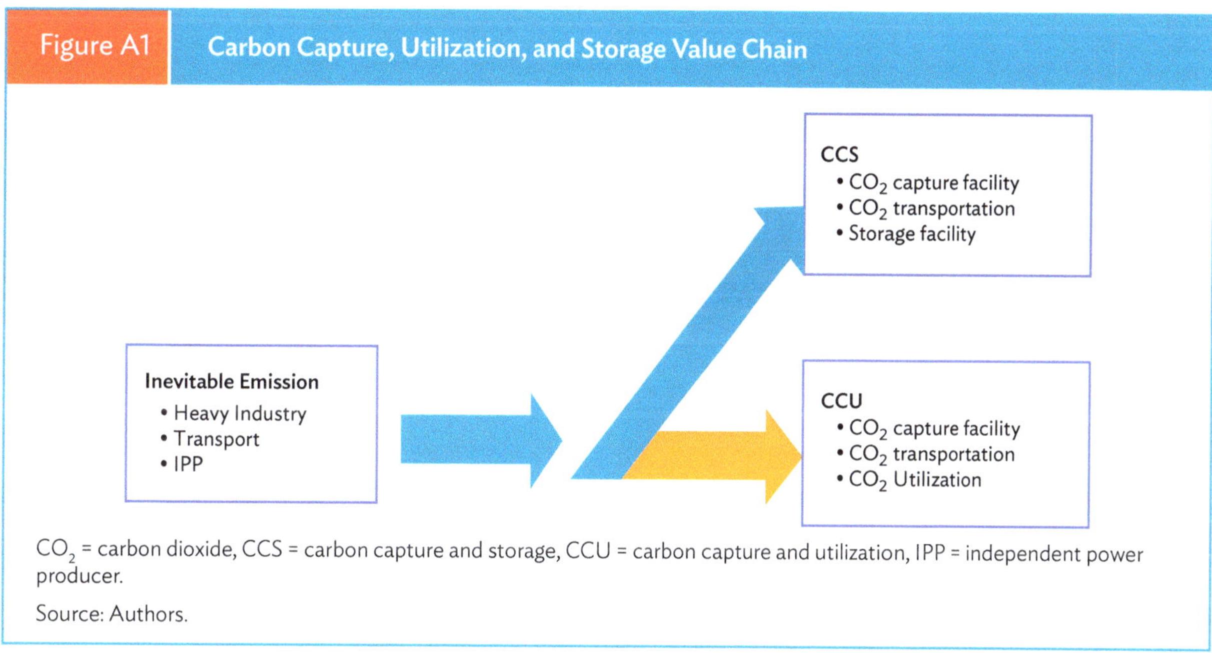

CO_2 = carbon dioxide, CCS = carbon capture and storage, CCU = carbon capture and utilization, IPP = independent power producer.

Source: Authors.

While the technologies are maturing, the most important factor affecting utilization is the cost associated with the various CCUS technology options. The key cost components of the CCUS operation are:

- CO_2 capture at the emissions source—purifying CO_2 from a gas stream up to over 95% purity by volume.
- CO_2 dehydration and compression/liquefaction, depending on the transport method.
- CO_2 transport by pipeline, ship, or mobile vehicle.

There is a considerable opportunity for the country to deploy CCUS in power plants and related industrial activity. This requires site-specific techno-economic feasibility studies, and the government has yet to introduce any policies or legislation focused solely on encouraging the development of CCS or CCUS. The storage and utilization of captured carbon is an evolving area, where several studies are presently ongoing globally to demonstrate the mitigation potential of the technology. One study has found the Titas gas field is found to be geologically the best gas field in the country for carbon sequestration (Hoque et al. 2019). However, considerable research on this needs to be conducted to aid future planning.

The overall technology readiness for different levels of the CCUS value chain has been reviewed by the Global CCS Institute, which finds that aspects of the carbon capture stage are relatively less developed than transport and storage (Table A1).

Table A1	Technology Readiness Level for Carbon Capture, Utilization, and Storage Value Chain	
Value Chain Level	**Technology**	**Technology Readiness Level Range**
Carbon Capture	Liquid solvent	2–9
	Solid adsorbent	1–9
	Membrane	2–9
	Solid looping	5–7
	Inherent CO_2 capture	5–7
CO_2 Transport	Compression	8–9
	Pipeline	8–9
	Truck	8–9
	Rail	7–9
	Ship design	3–9
	Ship infrastructure	2–9
CO_2 Storage	Depleted gas and oil field	5–8
	Saline formation	9
	Enhanced oil recovery	9
	Unconventional storage (basalt and ultra-mafic rocks)	2–6
	Unconventional storage (enhanced coal bed methane)	2–3

Note: a higher score indicates greater readiness.
Source: Global CCS Institute. 2021.

References

ADB. 2017. Improving Lives of Rural Communities Through Developing Small Hybrid Renewable Energy Systems, https://dx.doi.org/10.22617/TIM178811-2.

Bangladesh Power Development Board, n.d., *Annual Report 2019–20*. https://bpdb.portal.gov.bd/sites/default/files/files/bpdb.portal.gov.bd/annual_reports/7b792f67_bf50_4b3d_9bef_8f9b568005c9/2022-10-18-05-37-2ae06b4716a042c4b39d5cb54ef20ff4.pdf.

Bhattacharjee, A., 2021. Energy Efficiency and Green Building Pathways. New Age. 23 February. https://www.newagebd.net/article/130872/energy-efficiency-and-green-building-pathways.

Daily-sun, 2017. CPGCBL, Mitsui sign deal for 600MW LNG power plant, published on 8 November 2017, accessed online on 23 November 2024 at https://www.daily-sun.com/post/267171.

Global CCS Institute. 2021. *Global Status of CCS*. https://www.globalccsinstitute.com/wp-content/uploads/2021/10/2021-Global-Status-of-CCS-Report_Global_CCS_Institute.pdf.

Halder, P. K., Paul, N., Joardder, M. U. and Sarker, M., 2015. Energy scarcity and potential of renewable energy in Bangladesh. *Renewable and Sustainable Energy Reviews, 51*, pp. 1636–1649.

Hoque, S. M. S., Iqbal, M. A., Ahmed, S. I. and Islam, M. A., 2019. Suitability Analysis of the Potential Gas Fields of Bangladesh for Carbon Dioxide Sequestration and A Simulation Approach for Titas Gas Field. Journal of Scientific Research, 11(1).

IEA. 2019. CO2 Emissions from Fuel Combustion; Emissions Database for Global Atmospheric Research (accessed 3 June 2019).

Improving Lives of Rural Communities Through Developing Small Hybrid Renewable Energy Systems, 2017.

International Energy Agency (IEA) Data, Bangladesh. https://www.iea.org/countries/bangladesh. 2021.

International Institute of Applied Systems Analysis (IIASA). 2018. *SSP Database (Shared Socioeconomic Pathways)-Version 2.0* (accessed on 23 July 2022). https://tntcat.iiasa.ac.at/SspDb/dsd?Action=htmlpage&page=about.

International Renewable Energy Agency (IRENA). 2020. *Trends in Renewable Energy.* https://www.irena.org/Statistics/View-Data-by-Topic/Capacity-and-Generation/Statistics-Time-Series.

Islam, K. M., 2016. Municipal solid waste to energy generation in Bangladesh: possible scenarios to generate renewable electricity in Dhaka and Chittagong city. Journal of Renewable Energy, 2016.

Karim, M. E., Ridoan, K., Islam, Md. T., Firdaus, M. S., Nurul A. B., and Muhtazaruddin, Mohd. N. 2019. *Renewable Energy for Sustainable Growth and Development: An Evaluation of Law and Policy of Bangladesh*. Sustainability. doi:https://doi.org/10.3390/su11205774.

Malek, M. I., Hossain, M. M. and Sarkar, M. A. R., 2015. Production and utilization of natural gas in bangladesh. Proceedings of the 7th IMEC & 16th Annual Paper Meet, 2(03).

Ministry of Environment, Forest and Climate Change (MoEFCC). 2015. *Intended Nationally Determined Contributions* (accessed on 25 June 2020). https://www4.unfccc.int/sites/ndcstaging/PublishedDocuments/Bangladesh%20First/INDC_2015_of_Bangladesh.pdf.

Ministry of Environment, Forest and Climate Change (MoEFCC). 2021. *Nationally Determined Contributions 2021 Bangladesh* (Updated) (accessed on 22 June 2022). https://unfccc.int/sites/default/files/NDC/2022-06/NDC_submission_20210826revised.pdf.

Our World in Data. https://ourworldindata.org/grapher/installed-solar-pv-capacity.

Prateek, S., 2018. World Bank Approves $55 Million for Renewable Energy in Bangladesh. https://www.mercomindia.com/world-bank-loan-bangladesh-renewables. (accessed 16 April 2024).

Pranti, A. S., Iqubal, M. S., Saifullah, A. Z. A., and Ahmmed, Md K. 2013. *Current Energy Situation and Comparative Solar Power Possibility Analysis for Obtaining Sustainable Energy Security in South Asia. International Journal of Scientific and Technology Research.* 2(8). https://www.ijstr.org/final-print/aug2013/Current-Energy-Situation-And-Comparative-Solar-Power-Possibility-Analysis-For-Obtaining-Sustainable-Energy-Security-In-South-Asia.pdf. (accessed on 20 July 2020).

PricewaterhouseCoopers. 2018. *Transforming the Power Sector in Bangladesh.* https://www.pwc.in/assets/pdfs/industries/power-mining/executive-summary-pwc-bippa-report-on-transforming-the-power-sector-in-bangladesh/transforming-the-power-sector-in-bangladesh.pdf.

Riahi, K., Van Vuuren, D. P., Kriegler, E., Edmonds, J., O'neill, B. C., Fujimori, S., Bauer, N., Calvin, K., Dellink, R., Fricko, O. and Lutz, W. 2017. The Shared Socioeconomic Pathways and Their Energy, Land Use, and Greenhouse Gas Emissions Implications: An Overview. Global Environmental Change. https://www.sciencedirect.com/science/article/pii/S0959378016300681. (accessed on 30 Jul 2020).

SREDA. 2020. Bangladesh Experience in Buildings Energy Efficiency Building Energy Efficiency & Environment Rating (BEEER) System for Bangladesh. https://globalabc.org/sites/default/files/inline-files/Bangladesh%20Experience%20in%20Buildings.pdf.

Shetol, M.H., Rahman, M.M., Sarder, R., Hossain, M.I. and Riday, F.K. 2019. Present status of Bangladesh gas fields and future development: A review. Journal of Natural Gas Geoscience, 4(6), pp. 347–354.

Talukder, B. and Hipel, K.W., 2019. Energy efficiency of agricultural systems in the southwest coastal zone of Bangladesh. Ecological indicators, 98, pp. 641–648.

Teske, M. S. and Nagrath, K. T. 2019. 100% Renewable Energy for Bangladesh-Access for Renewable Energy for All within One Generation. https://www.worldfuturecouncil.org/wp-content/uploads/2019/10/100-Renewable-Energy-for-Bangladesh.pdf.

UN Environment Programme. 2018. Development of a certification course for energy managers and energy auditors of Bangladesh. https://www.ctc-n.org/calendar/events/development-certification-course-energy-managers-and-energy-auditors-bangladesh.

World Bank. 2021. World Bank Data, Bangladesh. https://data.worldbank.org/country/bangladesh. (accessed on 12 September 2021).

www.ingramcontent.com/pod-product-compliance
Lightning Source LLC
LaVergne TN
LVHW071450180726
843512LV00018B/1334